THE REAL GORBALS STORY

Colin MacFarlane pictured during a May Day parade in 1967, standing up, holding the pole beside then councillor Frank McElhone. (Photo © Joseph McKenzie)

The
Real
Gorbals
Story

True Tales from Glasgow's
Meanest Streets

Colin MacFarlane

MAINSTREAM
PUBLISHING

EDINBURGH AND LONDON

For my mother and father, Betty and Colin – real Gorbals people –
and my children, Philippa and James

First published in Great Britain in 2007 by
MAINSTREAM PUBLISHING COMPANY (EDINBURGH) LTD.
7 Albany Street
Edinburgh EH1 3UG

ISBN 9781845962074

All picture section photographs courtesy of Glasgow City Council Development
and Regeneration Services unless otherwise stated

The author has made every effort to clear all copyright permissions, but
where this has not been possible and amendments are required,
the publisher will be pleased to make any necessary
arrangements at the earliest opportunity

This book is a work of non-fiction based on the life, experiences and recollections
of the author. In some limited cases names of people, places, dates, sequences
or the detail of events have been changed. The author has stated to the publishers
that, except in such minor respects not affecting the substantial accuracy
of the work, the contents of this book are true

A catalogue record for this book is available
from the British Library

Typeset in Stone Print

Printed in Great Britain by
William Clowes Ltd, Beccles, Suffolk

ACKNOWLEDGEMENTS

I have come across a variety of great people while writing this book. My thanks go to Nerys Wyn Davies of BBC Wales's *Pobol y Cwm*, who encouraged me to get my stories in print; the lovely Lorraine Kelly, for providing the foreword; my brother, Ross MacFarlane, who provided the family photos; Ron Smith from Glasgow District Council's planning department, who was a real gentleman and came up with a plethora of Gorbals photographs; Bill Campbell, boss of Mainstream, who gave me the chance to get the book off the ground; and his team, including Graeme Blaikie, who handled everything with efficiency and professionalism, editor Claire Rose, who was brilliant at confronting the often bizarre subject matter, and Lee Fullarton, who designed the lovely front cover. I am grateful to my old pals Chris and Peter, whom I met up with in the Gorbals after all these years; Alan Cotton, ex-mayor of Pontypridd and and excellent IT advisor; Eric Talbot, former Glasgow-based archeology lecturer, who gave me great research material; my former editor at *Celtic Press*, Dudley Stephens; barman Dickie Mint, of Cardiff's Marriot Hotel, who had many great drinking tales; my auntie Kathleen Wright, who put me up in her flat in Govanhill when I was researching the book; and finally, the staff at the Mitchell Library, who were of immense help.

CONTENTS

FOREWORD

Reading *The Real Gorbals Story* was a journey back in time to my childhood. Anyone brought up in the Gorbals or with links to this extraordinary part of Glasgow will be completely enthralled. Everyone else will have their eyes opened to this colourful, funny, warm, frightening, dark and dangerous place.

After getting married in Martha Street registry office, my mum and dad set up home in a wee single end in Ballater Street and that's where I spent the first couple of years of my life, before we moved to Bridgeton and unimaginable luxury – a room and kitchen and an inside toilet.

My father was born in Mathieson Street and my granny Margaret Kelly still recalls the Gorbals as a real community, with doors being left open because no one had anything worth stealing. She brought up four kids – John, Billy, Lydia and Carol – in the Gorbals at a time of great hardship but with strong ties of friendship in the close where they lived.

My great-grandparents lived in the Gorbals all their lives and ended up stuck high in the sky in the late and very much unlamented Queen Elizabeth flats after their tenement was reduced to rubble in the clearances.

They would have loved this book. Colin brings the old Gorbals back to life and paints a vivid picture of the characters who inhabited the streets and pubs. I laughed out loud at the patter, the cockiness and the sheer sassy cheek of them. You can almost smell the New Year steak pie and dumpling, just like my granny still makes, as well as the close where the lobby dosser slept off his hangover.

This is an affectionate but honest account of growing up in what was a boozy, brash, brutal part of Glasgow with a worldwide reputation for poverty and the hardest of hard men. The famous Gorbals sense of humour shines through but Colin never spares us from the violence, crime and senseless brutality. It's a rich, rough raucous ride. Enjoy!

Lorraine Kelly
July 2007

PREFACE

This book is for all those Gorbals natives scattered all over the globe who still pine for the place. I was part of the last generation to grow up in the old Gorbals, during the 1960s, and witnessed its destruction at first hand.

My parents moved to the Gorbals in the 1950s and lived there until the early 1970s. They considered it to be one of the most exciting and interesting places to live in Scotland and perhaps the world. During that time, the Gorbals had a wonderful community atmosphere, full of life and vibrancy. It was like being part of one giant family as we dealt with life's everyday trials and tribulations. The humour was always there and the street patter could transform a wretched state of affairs into an entertaining experience. There was an abundance of colourful characters who were a delight to watch and hear. Some people tend to portray the Gorbals in a negative light but my overriding memories of growing up in the place are positive. It was a true melting pot where people from all over the world lived and worked together.

More than 30 years later, I still hunger for the old place and its former inhabitants. When I do take a trip down memory lane and visit the area, all I can see now are hundreds of modern luxury flats inhabited mostly by outsiders who know very little about the historical and cultural significance of the area. But times change and all good things come to an end.

It is just a pity that during the redevelopments of the '60s the authorities did not preserve the beautiful old buildings of the Gorbals instead of destroying them. When the buildings went, the people went as well. Now the former inhabitants are like rolling stones, wandering all over the world, perhaps searching for a place that can bring back the happiness of the good old days. Even now, when I dream, and meet people in my dreams, I am still on the streets of the Gorbals, outside the tenements, pubs and shops, with all the characters. Of course, the places and most of the people I dream about no longer exist. But they had such an influence on me that they are firmly carved into my subconscious. In this book, I didn't want simply to write a memoir of my childhood but to create a portrait of the time and place in which I grew up and for this reason some literary devices (for example, composite characters and imaginative reconstruction of dialogue) have been used.

If you are from the old Gorbals, you shouldn't feel miserable that it is no longer there. Just be happy that we all have wonderful – and sometimes astonishing or bizarre – memories of a time when we belonged to the Gorbals and the Gorbals belonged to us.

Colin MacFarlane
colinmacuk@yahoo.com

Chapter 1

MURDER POLIS

Tell me a story,
Tell me a story,
Tell me a story, before Ah go tae sleep.
Tell me aboot the birds and bees,
Tell me aboot the little trees,
Tell me a story, before Ah go tae sleep.

Gorbals nursery rhyme

Once upon a time in the Gorbals, there was a dilapidated public house called the Britannia. It was a seedy Victorian establishment, situated at the bottom of a grimy tenement block that had seen far better days. But the legendary Britannia was still very popular with the hundreds of heavy drinkers and hard cases who frequented the place. Known locally as 'the Brit', the bar was situated right in the heart of the notorious old Gorbals, on the corner of Thistle Street and Rutherglen Road. It had long been branded as a drinking shop that catered primarily for winos, bampots, headbangers and generally the element of society that most respectable people went out of their way to avoid. The Brit had become known for being the roughest, toughest drinking den in the whole of Glasgow and it had no problem at all living up to that dubious reputation. We used to joke that the pub was so rough that anyone in there without a scar was probably a tourist. It was even said that the Brit was so intimidating the police rarely ventured inside because they were too afraid of what might happen in such an unpredictable environment.

When we Gorbals street boys looked through the smoke-stained windows of the Brit, the scene resembled not so much something you might expect from the so-called swinging '60s but rather a scene from the impoverished '30s. The Brit's vast cast of dodgy characters looked like they had stepped straight out of the book *No Mean City*. We often imagined that fictional razor king Johnnie Stark might walk through the doors any minute. I was born in late December 1955, more than 20 years after *No Mean City* was published, and lived in Crown Street, where Stark's razor-slashing antics were based. Two decades later, many of the character types described in the book were still around and the Britannia was living testimony to that. The hard drinking and open-razor fights still went on and the area, with its magnificent buildings and crumbling tenements, was much the same as it had been in the 1930s. The men continued to wear bunnets and played pitch and toss in the street, women still wore headscarves while pushing cumbersome-looking prams through the streets and families numbering as many as 12 lived, slept, ate and washed in a tiny single end. Rats ran around the place, the street gangs continued to run amok, down-and-outs guzzled from wine bottles on the corners, crime was rife and the majority of Gorbals people still seemed resigned to living in conditions that resembled something from a Third World country. Glasgow was once considered the second great city of the Empire yet the stench of human waste, which often ran down tenement stairs, was always in the air.

Every night at the Brit, the Gorbals characters got stuck right intae the bevy, drinking pints of foaming McEwans heavy, hawfs o' Bells whisky and the two popular, and very powerful, South African wines El Dorado and Lanliq. 'Ah love ma cheap wine,' Auld Tam, an experienced bevy merchant, confessed to us. 'It gives ye a nice wee lift and it takes yir mind aff aw yir troubles. But the problem is wan minute ye're laughin', the next minute ye're greetin'. It's funny how the wine affects ye but ye get used tae it after a while. A wee glass o' wine dis ye the world o' good. Even the Bible will tell ye that – give wine tae a troubled soul, it says.' There were certainly plenty of troubled souls around the place.

It was the cheap wine that had the Brit rabble singing, shouting and bawling, crying, laughing and occasionally fighting. The language used on such occasions was, you might say, unsophisticated. 'Ya f****n' dirty bastard, ye, Ah'll tear yir heid aff wi ma bare hauns and then Ah'll throw it tae the rats,' we heard one wine-sodden man shout to another. But circumstances were always unpredictable and the next minute everyone might be neighbourly again, aw pals at the palace, shaking hands and hugging each other.

Then they might launch into one of their songs, the favourite at the time being the country-and-western song 'The Wild Side of Life', which usually sounded way out of tune but on other occasions was not bad at all. Somebody would shout, 'How's aboot a wee go at "The Wild Side o' Life"? Aw the gither noo . . .' And we'd see the tears welling up in some people's eyes as they began singing. The wine-fuelled camaraderie made everyone very emotional as they thought about their often short, austere lives in the Gorbals.

My uncle Mick had never been in the Britannia when he moved to the Gorbals. He had come from a tough area of Clydebank and thought he was a real man about town, no mug, until he walked through its doors. He was a Glasgow Corporation bus driver and, it being Friday night, had just had his pay packet. He put a crisp Bank of Scotland pound note on the counter and ordered a pint of heavy and a large glass of El Dorado, otherwise known as 'El D'. But then a big scar-faced fellow came over, reeking heavily of cheap wine. He stuck his large hand over the note and declared, 'That pound's mine, pal, so get tae f***, unless ye want tae end up in the Royal Infirmary.'

Mick told me, 'This guy had so many scars on his face it looked like wan o' they road maps Ah use when Ah'm tryin' tae work oot a new bus route. Ma first instinct wis tae stick the heid oan him or kick him in the baws, but he flashed an open razor in ma face, then near ma throat, and Ah could see jist by lookin' aroon' he wis too well-handed. He wis wi a gang o' ugly guys wi scars and broken noses, aw stinkin' o' wine. So Ah decided tae dae the best thing, offski oot the place leavin' ma pound behind. It's the nearest thing Ah've ever seen tae a lunatic asylum and in that place the lunatics are running the asylum. Ah hid tae let ma money go. A bampot like that wid slash yir throat fur a drink.'

The Gorbals had a reputation for its toughness not only throughout Glasgow and the rest of Scotland but all over the world. *No Mean City*, by unemployed Gorbalian Alexander McArthur and English journalist

Kingsley Long, portrayed the heavy drinking, gang fights and poverty, and put the Gorbals in the spotlight worldwide. As a result, in the years following the book's publication the media always concentrated on negative images of the area. Drinking dens like the Britannia boosted that perception but if you lived there, you knew that behind the myth the whole of the Gorbals wasn't like that. The national and international bad impression of the area often overshadowed the positive reality of living there. Most of the inhabitants were clean, honest, friendly, non-violent people and there was always a great deal of friendship and humour running alongside the hardships of everyday life.

As in other working-class districts in Glasgow, people like myself lived up a close. This was usually a three-storey Victorian tenement building with up to six flats on each of the floors. The closes bred their own lifestyle, even down to the basic duties of keeping them clean. For example, every week it was some wee woman's turn to do the stairs. These housewives had no hesitation in getting on their hands and knees to scrub the tenement steps with strong disinfectant, making them spotless. They took great pride in keeping a clean close. When we watched them, it was as if they were scrubbing away the worries and failed dreams in their lives.

As they did so, they often sang Glasgow street songs which they had learned playing on the streets when they were girls. One of my tenement neighbours in Crown Street, called Wee Betty, regularly got on her hands and knees and happily sang the song 'Hairy Mary' as she cleaned the close:

> Ah'm no hairy Mary, Ah'm yir maw,
> Ah'm no hairy Mary, Ah'm yir maw,
> Ah'm no hairy Mary, Ah'm yir maw's canary,
> Ah'm no hairy Mary, Ah'm yir maw.

Most of the local people were hard grafters and would take up any employment, usually poorly paid, just to keep themselves occupied and out of mischief. 'See, the thing is,' said Jimmy, a hard-working labourer, to us, 'ye hiv got tae work, even if it is only fur a couple o' bob, cause if ye don't, ye end up drinkin' and landin' yirsel in a predicament and that's nae good fur any man. A lot o' people who gie up work end up as nae good wine-moppers. Ah don't want that tae happen tae me. Ah've got a wife and two weans tae look after. The devil makes work fur idle hauns. Trouble always happens when ye're no workin' – jist look at they imbeciles in the Britannia.'

To us Gorbals boys, watching the characters of the Britannia and hearing their patter was always an entertaining experience but to visitors from outside the area, a trip to the pub often left them shell-shocked. 'Gaun tae the Britannia is the best laxative some people could ever hiv. At times it really wid scare the shite oot ye,' Jimmy maintained. He was fairly accurate: the Brit boozers were the sort of human beings who had given up any attempt to lead a wholesome life. They had become the people of the abyss, forced into their miserable existences by years of living on the breadline, unemployment and pishing their money up against the wall.

Most nights, a drama unfolded. Sometimes it was pure pantomime. At other times it was what some people aptly described as 'a real horror picture, full o' nutcases tryin' tae kick each other's heids in'. Drunken men battered the living daylights out of one another, with blood flying everywhere, often convinced that they were fighting for the championship of the world. Sure, world flyweight champion Benny Lynch had come from the Gorbals but he had succumbed to the bottle himself when his career ended.

We realised it was the low-priced wine that fired their liveliness and powered their aggression until they felt like invincible Gorbals gladiators. Most of them were acknowledged winos, on the elixir as soon as they got out of their beds in the morning. The fights usually lasted only a few minutes or even seconds, as the men were often too intoxicated to stand up properly. There was always the usual effing and blinding, followed by attempts at head-butting, kicking and punching. 'Ah'm gonnae kill ye,' we heard one inebriated man shout as he proceeded to have a fight in the middle of Thistle Street. He was a small man, barely bigger than a midget, but his voice had the threatening air of a violent giant. 'Ah'm gonnae teach ye a lesson that ye'll never forget. Mess wi me and ye're messin' wi the wrang people, ya tube, ye.'

People who lived in the tenements opposite in Thistle Street often did a bit of windae hingin', putting a nice plump pillow under their elbows to catch the night's action. They considered it far more entertaining and stimulating than the cinema. To these windae hingers, the Britannia was the Gorbals' real-life action movie, full of melodramas that few screenwriters could have concocted. 'Hey, who needs the pictures when ye've got the Britannia Bar? It's got mair action and mair goodies and baddies than a cowboy western,' one of the windae hingers joked to us.

As the men were punching the stuffing out of each other, money flew out from their pockets and cascaded all over the street. To the local kids, this provided a much-desired windfall to improve their impoverished lifestyle. Some nights, especially during the summer, us youngsters could make a couple of quid just by standing around and watching the battles taking place. Wee Alex, a ten-year-old pal of mine who was streetwise beyond his years, came up with an ingenious idea. When two men stepped out of the pub and headed to have a square go in the adjacent spare ground, Alex would walk up and say, 'Hey, Mister, can Ah haud yir jackets?'

Usually the drunks obliged and handed Alex their jackets to hold. Nine times out of ten they contained significant amounts of small change and even pound notes. When the men started hitting each other, Alex ran off with the jackets to a nearby close and helped himself to their money. He then rushed back to the battle as it ended and handed the bloodied contestants their garments. The fighters were usually too stunned, concussed or inebriated to notice that not only had they both been battered but they'd been robbed as well. Alex once said to me afterwards, 'Whit a pair o' bampots. There they are knockin' the f*** oot o' each other and Ah've got aw thir money. Some people will jist never learn, will they? Fifteen bob the night, that wisnae bad, wis it? Better than workin' fur a livin'! There's a fool born every minute, especially a drunken wan.'

Alex was always coming up with ideas to make money – it was imperative. His mother had brought him up single-handedly, along with his brother and sisters, in a cramped, damp single end in Thistle Street on little or no money. It was simple: he had to continually come up with ideas to survive, by hook or by crook, otherwise he and his family would go without.

Every Friday and Saturday night, hundreds of men staggered along the streets of the Gorbals but very few actually came to a sticky end. They all seemed to have a survival instinct, like homing pigeons. It ultimately kept them out of harm's way and ensured they would be safely back home in their beds at the end of the night – although there were exceptions to the rule.

One Friday night, a drunk man left a pub in Crown Street and then, as he floundered across the road, a speeding car hit him full force, sending him flying into the air. To us, it was as if the whole scene were being played out in slow motion. As he flew into the air, we could see his face twist and contort before he fell dying onto the pavement. An ambulance was called and a crowd gathered round. We watched in shock as dark-red blood began to flow from his mouth. It was a disturbing sight. One woman held his hand tightly and kept saying, 'Ye'll be awright, son, ye'll be awright.' But it was clear to all around that she was giving out false hope. The man's eyes rolled like a pair of marbles and he mumbled, 'Tell ma wife it wisnae ma fault.' By the time the ambulance arrived, he was dead. Some of the women in the crowd that had assembled began shrieking, 'Oh, it's a bloody shame, so it is! Bloody shame, such a young fella, he shouldn't have died like this.' Police later identified the man after finding a pawn ticket in his trouser pocket. It turned out he had pawned his Sunday suit so that he could have a drink that night.

During the 1960s, there was an influx of Irish labourers to the Gorbals. They always had plenty of money to get steaming, especially on a Friday night when they'd just been handed their bulging pay packets on the building sites. They meandered around the streets still in their mud-stained wellingtons and donkey jackets. We could hear the money chinking in their pockets as they passed by. It was like alarm bells going off. Me, Alex and the rest of the boys reckoned this would be a great source of extra income, so we decided to start dipping them – picking their pockets. Alex said: 'Wi hundreds o' navvies staggerin' through the streets, it's gonnae be like wan big game o' skittles. Wan o' them will fall doon eventually. That's when we move in.'

He was right. A drunken man was often to be discovered lying on the pavement and we would help get the unfortunate individual back on his feet again. Of course, we always charged a fee by dipping the man. The proceeds were usually what was left of his wages after the barmen and bookmakers had taken their cut. Every weekend, the same scenario was played out. Some nights, the dipping business was booming and we 'helped' up to 20 men.

Friday night had its usual parade of characters. Big Paddy Malone, a giant of a man from Donegal, had hands like shovels and drank like a fish. He was no mug and even the local gang members were wary of tackling him. One night, we witnessed him knocking out three fellow Irishmen in a fight over a building-site disagreement. It was like watching Goliath in action. He boasted afterwards, 'Serves them right. They should have known that I'm rough, tough and hard to bluff.' We would never have gone near him when he was sober but when the drink took its toll he was, as wee Alex said, like a big dod of putty in our hands. Paddy often stood at the bar with a pint of Guinness in one hand and an Old Bushmills whisky in the other. As he got more inebriated, he would shout a rhyme he had learned during his drinking days in Ireland:

> There's a gladness in my gladness when I'm glad,
> There's a sadness in my sadness when I'm sad,
> There's a madness in my madness when I'm mad.
> But the gladness in my gladness,
> The sadness in my sadness
> And the madness in my madness,
> Is nothin' compared tae the
> Badness in my badness when I'm bad.

He and his compatriots would then launch into a sing-song, which was usually 'Off to Dublin in the Green'.

One summer's night at Derry Treanor's pub in Gorbals Street, Paddy and his compatriots embarked on an innovative way to get smashed quickly. They were all having a ferocious dispute, with the big man leading the way. He boasted, 'Ah could drink any o' ye under the table. There isnae a man alive in Ireland, Donegal tae Dublin, that can drink a bottle o' whisky faster than Paddy Malone. I'll bet any fella here that he cannae beat me.'

We looked through the pub windows as Paddy arranged for two bottles of Old Bushmills to be put on the bar. He then challenged a younger fellow from County Cork to drink a bottle down in one go. There was a twenty-pound bet on it and the whole pub looked on as Paddy and his pal unscrewed their bottles. The contents of Paddy's bottle went down his throat in a matter of seconds, as if it were Irn Bru. The Cork fellow was slightly slower but still managed to down his whisky only a few seconds after Paddy had finished. The next minute, Paddy's eyes began to bulge and he staggered over to the bar to pick up his winnings. He was shouting, 'Beat ye, ya eejit, I told ye nobody but nobody kin beat Paddy Malone.' The young Cork man, who couldn't have been more than 19, began whimpering like a child, 'Oh, Mammy, oh, Mammy, help me,' before collapsing onto the pub's floor.

Paddy lurched out of the bar and staggered several hundred yards before eventually collapsing. It was like watching a dinosaur fall. We went over to help him back on his feet but found it impossible because of his immense size and his whisky-sodden condition. Summoning all our strength, we eventually managed to drag him a few feet to a nearby tenement wall and prop him up. A few minutes later, Alex was clutching the twenty pounds plus change from Paddy's donkey jacket. He was elated, shouting, 'Two big daft Irishmen hiv tae drink a bottle o' whisky doon in wan go and we get aw the money! It's no a bad game, is it? It's true whit they say: whisky wis invented so the Irish will never hiv a chance o' ruling the world!'

A local patter merchant who had been watching Paddy's exploits said that he had thought up a new drinking game: 'Ye put a bottle of whisky oan the bar and stick a blindfold oan. Ye drink the bottle doon in wan go and take the blindfold aff. Then ye've got tae guess who ye are.' Strong drink was always a subject for jest and comment. A man who was intoxicated might be said to be 'bevied', 'miroculous', 'jaked', 'paralytic' or 'stovin''.

Unlike nowadays, we rarely spotted a woman going into a pub by herself. It certainly wasn't the done thing in the Gorbals. Sections of the pubs were set aside for women and their partners to get together and have a drink. The public bars were always full of men. You would find the ladies in the lounge with their partners at the weekends. A lot of relationships started in such lounges, because people often got so drunk they literally fell into each other's arms. The courting rituals were interesting to watch, with a romantic saying lovingly to his girlfriend, 'Ah love ye, doll. Here's my haun, here's ma heart.' Drink, once again, had a lot to answer for.

The worst sight to see was that of a drunken woman staggering along the street. We never ever contemplated dipping a woman, as she was somebody's mother, wife, sister or granny, and we usually gave her a hand to get back home. Some of the drunken women were sorry sights. One such, Annie, who had recently moved into the area, lived up my close. I initially thought of her as a respectable type who had flitted from another part of Glasgow for some mysterious reason.

She was a big woman in her late 30s, with long red hair and two young children. She told my mother, 'Ah broke up wi ma man because he wis beatin' me up when he hid a drink in him. He's nae good tae man, woman or beast, an Ah jist hope when he comes oot o' jail he disnae try tae track me doon tae the Gorbals.'

I spotted Annie going into the Britannia one Friday night. I couldn't believe my eyes. Neither could the rest of the boys, because no respectable woman would frequent such a place on her own. Except, as my pal Albert explained, 'winos and dirty auld cows'. Albert and I watched her through the ajar front door; she was drinking large glasses of El D with a seedy-looking man who had a small moustache. He was in his late 30s and had an abundance of dough on him. He was comparatively well dressed, with all the characteristics of a pimp. Albert recognised him as a guy called Charlie he had seen with two prostitutes at Gorbals Cross the week before. Albert said Charlie was a low-life bampot trying to pretend he was big-time, a big shot. He was right. Charlie was throwing his money about, buying Annie round after round of cheap wine. He was also giving her all the 'lovey-dovey' patter: 'Awright, darlin'? Drink as much as ye want, Ah've got enough money on me tae choke a donkey. In fact, it wid choke quite a few donkeys.' They both hee-hawed at his witticism; he was indeed making a real Charlie of himself.

When the pub closed at nine o'clock, they stumbled out into Thistle Street arm in arm. Annie was laughing and giggling, and there was a lot more of the drunken lovey-dovey gibberish. She was saying, 'Ye know, Charlie, Ah've always hid a thing fur ye. The day his jist been dead brilliant, being wi ye. Ye're far better than that big chancer Ah wis married tae.' They headed to a close just across the road. Myself, Albert, Alex and a few of the other boys followed them quietly into the darkness and could hear loud grunting and moaning as they began to make love. We had become used to such sounds and sights. After the pubs closed, couples with nowhere to go for privacy would make love in the dark, dingy closes or manky middens. The guy's trousers were around his ankles and his bare behind shone like a lamp in the moonlight. Alex picked up a large stone and threw it at Charlie's behind, which made him jump into the air in pain and fright. He screamed, 'Oh ya! Oh ya! Ya dirty wee bastards,' pulled up his trousers and tried to chase us. But he had had too much to drink. Annie came out of the close screaming hysterically at us, 'How dare ye! How dare ye! I'll get ye fur this!' But we did not feel at all afraid. Tougher people had threatened us. We considered it a joke that Charlie had called us dirty wee bastards when, as Albert commented, he was a dirty wee bastard himself. Annie, with her flaming red hair all over the place, was bawling her head off and continued to threaten us, screaming obscenities. But she was horrified when she saw me, because the wine was wearing off and she recognised my face. She knew that I knew her, that I was her neighbour.

The next day, I was going up the tenement stairs when I saw Annie, now completely sober, coming towards me. She didn't say a word and just passed me by. But I gave her the sort of look that said, 'Ah know whit you've been up tae.' In that brief flash of time, she realised that her secret was out and a look of terror and despair swept over her face. It was like watching a rabbit frozen in the headlights of a car.

A few hours later, I heard a loud siren and looked out of the window to see what was going on. There were two police cars, an ambulance and hundreds of people outside our tenement. I ran down the stairs and joined the throng. I asked one of the women in the crowd what had happened and she shook her head saying, 'The daft bastard Annie jist cracked up. She swallied a bottle o' aspirins and then threw hersel oot the windae. It's the poor wee weans Ah feel sorry fur. Ah mean, imagine hivin' a maw like that? She couldnae hide the fact that she's a loose woman. Bloody disgrace it is.' Other people in the crowd nodded their heads in agreement. A stretcher carried the woman away and a pair of grim-faced social workers took her children into care. Annie never came back to our close again but she did recover enough to move to another part of Glasgow, the Cowcaddens, presumably to reinvent herself. For me, at the age of ten, it had been an important lesson in human nature and behaviour, one of the many I would learn on the streets of the Gorbals.

Another of our neighbours was a woman called Rosie, a big chatterbox and a teetotaller who lived just across from the Britannia. Rosie, who was a little woman in her early 60s, loved her bingo with a passion and had a great sense of humour. Every night, we shouted to her, 'How's it gaun, Rosie?' She usually replied with a wry smile and a shrug, 'It's murder polis livin' across fae the Britannia. They should shut that bloody place doon. It's full o' heidcases and hingin' is too good fur them.'

One summer's night, we had just said hello to Rosie and twenty of us were having a kick-about with a two-and-a-tanner football on the spare ground next to the Britannia. It was just past eight o'clock and still light when two men came out of the bar. The first fellow was a wiry-looking man in his 30s, wearing a tartan bunnet and heavy-rimmed glasses. The other man was older, in his 50s, and had the stout build of a serious beer drinker. Both of them had obviously been drinking all day. When they came out of the Britannia, they were laughing at some joke or other. But then the mood changed, with the man in the tartan bunnet shouting, 'Where's ma f****n' money? Ah warned ye tae get it fur me the night otherwise there wid be trouble. Ye've hid long enough to pay me back. Don't even think o' tryin' to bump me. Ah didnae come up the Clyde on a banana boat, ya f****n' doughball.'

The stout fellow, whom we had nicknamed 'Bawheid', looked a bit taken aback at first but the cheap wine had given him courage and he pushed away the guy with the tartan bunnet, shouting back, 'Ye'll get yir money awright! Don't threaten me! Ye're jist a wee bampot, I could take you any time, so get tae f***. Who dae ye think ye ur? It's been me that's been buyin' ye drink aw day. You've no put yir haun in yir pocket wance, ya miserable bastard, ye.'

The argument continued and the language got worse. More insults were traded, then Bawheid pushed Tartan Bunnet yet again and they began grappling with each other. I thought it looked pathetic at first, more like a schoolyard scuffle than a real fight. But then Tartan Bunnet took a couple of heavy blows to the face and, smeared with blood, he wobbled like a big jelly. It looked as though he was going to hit the pavement. Suddenly, he pulled out an open razor and slashed Bawheid full across the throat. Bawheid just stood there, holding his slashed throat in a state of deathly disbelief. After a few seconds, all the colour drained from his face and the blood gushed into the air like a fountain. Looking back, I'm sure the main artery must have been cut. He staggered a few feet, gurgled, then collapsed in the middle of road. His blood flowed all over Thistle Street. Tartan Bunnet sobered up with the shock of what he had done and shouted to us, 'If anybody asks ye whit happened, ye saw f*** all.' He then ran off through the back courts as if he was a rabbit being pursued by a greyhound.

Not long afterwards, the place was swarming with police and although plenty of people had seen the murder unfold, including ourselves, not one witness came forward. It wasn't the done thing to talk to the police and if anyone did, no matter what the crime was, they would be branded a grass. A policeman with a Highland accent came over to us and asked, 'Did ye see anything, boys? Ye must have, cause ye're always playin' fitba here, day and night, and there's no much that goes by that ye don't know aboot. Sure, you guys at times are the eyes and ears o' this part o' the Gorbals.'

One of the boys, Chris, shouted with an air of contempt, 'Aye, Ah know who it wis, it wis the man wi the bunnet that done it.' We all laughed, because that phrase had become a standing joke when we were trying to ridicule the police. Ironically, Chris was telling the truth – the man with the bunnet had done it – and the policeman would never realise it. The policeman's face filled with revulsion and anger, and he began shouting, 'Think ye're a lot o' wee fly men, dae ye? Jist ye wait, I'll be back in the no too distant future tae teach ye aw a lesson. Ye'll no be laughin' when ye're aw locked up fur a very long time.' We Gorbals boys, all witnesses to the murder, just laughed it off. We reckoned that it was his job to catch the reprobate and he was being paid good money to do it. Besides, the unwritten rule of the Gorbals was that we couldn't and wouldn't help him. Later on, we read in the Glasgow *Evening Times* that Tartan Bunnet had been arrested in connection with the murder. It turned out that the murdered man had been on the run from the police after a bank robbery and Tartan Bunnet was a well-known moneylender.

Before they took Bawheid's body away, the police drew white chalk marks around it and they were still there the next morning when we went out to play. A few lassies joined us and pointed to the chalk marks. One of them said it was an ideal spot to play peevers. So there we were, Gorbals boys and girls on a bright summer morning hopping around and playing on the chalked image of a dead man. One of the lassies began to sing:

> Murder, murder polis, three stairs up
> The woman oan the middle flair hurt me wi a cup.
> Ma heid's aw bleedin' and ma face is cut,
> Murder, murder polis, three stairs up.

I realised then that Rosie had been right: it really was murder polis living across from the Britannia.

Chapter 2

A GORBALS FAMILY

All sorts of people ended up in the Gorbals for all sorts of reasons. My family were no different. My mother, Betty (née Crumlish), often got tearful and recalled how her mother had died at an early age leaving her, two brothers and a sister to be brought up in a care home while their father was away fighting in the Second World War. When he came back to Glasgow, her unhappy childhood continued, with a stepmother taking over the care of the family. They moved to a tenement in Garscube Road in the Maryhill part of Glasgow and my mother's greatest ambition then was to escape as soon as possible. Later, the family moved to the Househillwood area. It was a dramatic change from the slum dwelling she had experienced in the decaying tenement buildings of Maryhill. In this new 'overspill area' of Glasgow, built on a green area half an hour's bus ride from the city centre, there were modern houses with two or three bedrooms, fitted kitchens and inside bathrooms. By the time she was a teenager, my mother had film-star looks and when I see pictures of her with long brown curly hair and wearing a string of pearls, I think she looked very much like Ava Gardner.

My father, Colin MacFarlane, had been brought up in another slum area of Glasgow – Partick. As an adult, he said that he had blotted out his childhood because of the experiences he had been through. He claimed that what had hindered him in life was the fact that when he went to St Peter's Primary School in Partick, he couldn't see the blackboard. Like a lot of children at the time, he was too embarrassed to admit to the teacher that he needed a pair of glasses but his parents couldn't afford to buy them.

He said he learned hardly anything at school but quickly became streetwise and that, coupled with his ability to walk and talk like a tearaway, helped him survive. He wasn't a big man, he was of medium build and height, but he knew how to handle himself, leading him to land in trouble during his youth. When his family also moved to Househillwood, he missed the vibrancy, aggression and humour he had experienced in Partick. In his teens, he became a member of a local clique calling themselves 'the Chain Gang'. He and about 20 other untamed youths roamed the district getting into various skirmishes.

My father also admitted that he had been involved in a lot of gang warfare, which had resulted in his being framed for assaulting another gang member during a street fight. It caused his family a great deal of distress, because he was sentenced to a short time in Barlinnie. But he wasn't considered a bad person and anyone who knew him said he was just a product of his times and wee bit wild for his age.

At parties in the Gorbals, he often sang a Glasgow street song that reminded him of his riotous youth:

> My pal got lifted last night when the lights were low,
> My pal got lifted last night, and it was off to Barlinnie he did go.
> He was charged with breach of the peace,
> For trying to disturb all the police.
> The judge said, 'One guinea, or ten days in Barlinnie,'
> Oh, pal, I miss you tonight!

Later, he joined the merchant navy as a cook and travelled all over the world, bringing back hundreds of gifts for his mother and father in Glasgow. My granny even bought a china cabinet to house the exotic presents from overseas. It was on a visit home that he fell for my mother and they began courting. Even then, in the 1950s, my father wore sharp suits and shirts, handmade for him on his trips to places like Hong Kong. He stood out from the crowd, as did my mother, and people said they were a smashing-looking couple. My father was still a bit boisterous, though, and fell out with my mother's father, with the squabble resulting in my dad putting a brick through his window. My parents ran away together, got married at Martha Street Registry Office in Glasgow and then fled to London. They planned to live there briefly, to make enough money to come back to Glasgow and buy a place to settle down in. They found it easy to get employment in the big metropolis: my father worked in a bakery, making bread, and my mother as a waitress in a restaurant. After more than a year, armed with their savings, my father and pregnant mother decided to move back to Glasgow.

It was 1955 and the Gorbals at the time was a thriving place, full of life and dynamism. They had £100 between them to get a house. They must have been naive, because bungalows not far away in a more respectable area were being sold for around £500. But my parents handed over the £100 to a woman at 134 Crown Street for 'key money'. This meant that they hadn't actually bought the place but had simply acquired the tenancy and they had to pay rent to a factor in Ballater Street every week. By Gorbals standards, the flat in Crown Street was pretty grand. It was two rooms and a kitchen on the top floor of a three-storey tenement in a close shared with eleven other families and a dentist's practice. At ground level, there was a bank on the corner and Angus the Chemist's, adjoined by other businesses such as sweet shops, tobacconists, bakeries, newsagents, pubs, betting shops and grocers.

I was born into this environment in December 1955 and my brother Ross followed 18 months later. In some ways, we were fortunate, because we were brought up in the Gorbals of the late '50s and '60s, and at that time, for all its drawbacks, it was a stimulating place to be. Now, reflecting on those days, my brother says growing up in the Gorbals was a completely different experience for the two of us. He basically disliked it; he says that, for him, it was 'like going through an assault course every day', while I tended to treat the place like an adventure playground.

Although people lived in terrible housing conditions near by, the Crown Street tenements were comparatively posh, having been built for the Victorian middle classes. We even had an inside toilet but this was nothing to swank about, as it was no more than a claustrophobic hole in the wall with a cistern. In most of the Gorbals, it was still necessary for several families to share an outside toilet. As late as 1950, 40 per cent of the houses in the Gorbals were over 100 years old and 60 per cent had no inside toilet. The outside toilets were dismal, insanitary places and in the winter they were always freezing cold. If it was very cold, tenement inhabitants used chanties, which they stored under their beds. These bedpans were then slopped out in the morning. For toilet paper, strips of the *Daily Record, Evening Times* or the *Glasgow Citizen* were used. Men on the street corners often had long discussions about what newspaper made the best toilet paper. 'Ah'm tellin' ye, the *Citizen* is the best because ye can cut it intae nice long strips,' I heard one man tell his cronies. The outside toilets were frightening places for children, who believed that various ghosts hung about there, especially when it was dark. But these 'ghosts' were probably drunk men staggering about in the shadows.

The washing facilities in the house were primitive. Because there was no bath, adults and children bathed themselves in the kitchen sink, in cold water, or they would boil a kettle on the gas hob for the luxury of having some hot water to wash in. This began to change in the mid-1960s. People bought 'on tick' a little electric water heater to put above the sink so they were able to have instant hot water. Carbolic soap was applied and everyone in the family usually had their own face cloth. This was a flannel that you used to clean yourself with, then rinsed it out and hung it up to dry for the next day. Children who did not wash regularly were subjected to cruel taunts when they got to school, being branded clatty, mingin', manky or boggin'.

A bath was a luxury. If anyone wanted one, they went to the Corporation bathhouse in Gorbals Street, where a steaming hot bath cost a shilling. The one bath of the week was often kept to the weekends. On a Friday night, dust- and mud-covered labourers lined up in a long queue to get into one of the bath cubicles. Later, they emerged all spick and span, and headed to the pub. It was gratifying sitting in the hot soapy suds as the grime and dirt of the Gorbals was stripped off. It made people feel cleaner and less worried about the conditions they lived in. There was never any scarcity of hot water at the bathhouse. If someone spent too long in the bath and there was a queue building up outside, the attendant knocked threateningly on the door and shouted,

'That's it, time's up! There's other people waitin', ye know, and they've no got aw day.'

The Gorbals baths also had a decent-sized swimming pool with hot showers and small dressing cubicles. At one point these swimming baths were the most popular in Glasgow, with thousands using them every week. Many of the men and boys, instead of going for a bath, had a swim then stood for hours in the shower room indulging in conversation. The topics ranged from politics to football to almost anything under the sun. Many of the men were self-educated and had spent hours reading books from the Gorbals Library on subjects like history, politics, philosophy and economics. As we stood under the showers, a retired guy called Donald, an ex-union official and self-confessed bookworm, even tried to explain Einstein's theory of relativity to us. 'Einstein wis a genius but even he couldnae work it oot,' he said. 'Naebody knows where time comes fae and naebody knows where it goes, but it's aw tae dae wi planets spinnin' through space. Ah've read dozens o' books on it an it's only noo Ah know the secret o' life.'

'Whit's the secret o' life?' we asked, as the hot shower water cascaded over us. 'The secret o' life is tae enjoy the passage o' time. Anybody kin dae it, it's that simple. We're only here fur a while, jist passin' through, so enjoy the journey.'

The power in the flats was paid for by putting a shilling in the meter. In fact, a shilling piece was used to pay for almost everything, including gas and electricity. Even the television was coin-operated in some households. There was always somebody looking for a shilling to get their power supply on the go again. 'Excuse me, son,' they would say, 'ye widnae hiv a spare bob on ye by any chance?' There was no question of houses having central heating. During the winter, there was often ice on the tenement windows, inside and outside, and people walked around shivering and uttering the familiar comic phrase 'It wid freeze the baws aff a brass monkey'. So to combat the wintry cold, piles of blankets and old army greatcoats were put on the bed, which inevitably caused people to sweat profusely as they slumbered. Extreme cold called for a hot water bottle but some people could not even afford that, so poorer families used an old ginger bottle filled with hot water and then wrapped in a piece of cloth. Mothers knitted balaclavas, scarves and woollen mittens to keep their children warm. They became part of the Gorbals street uniform. The woollen mitts were joined together by a long piece of string. It went up one sleeve, around the back and down into the other sleeve, so the child would not lose them.

Pest control was an essential part of everyday living and mousetraps and rat poison were regularly put down to ward off the vermin. If there were fleas in the blankets, houses often had the repulsive smell of paraffin, which was used to kill them. 'Ah killed a flea this mornin', because it began tae irritate me when Ah wis lyin' in bed,' joked a neighbour to me, 'but it wis the 10,000 that turned up for the funeral that really annoyed me.' During the summer, flies were all over the place, with bluebottles being the main nuisance. Children like myself used rolled-up newspapers and spent all day in gangs swatting them. The winner was the swatter who had killed the most flies. One summer afternoon, we began a fly-swatting competition in the back courts close to the overflowing middens. I was announced the winner after accumulating more than 200 successful swats.

The appalling living conditions in the Gorbals first started to grab worldwide attention after *No Mean City* was published. In 1948, the mass-circulation *Picture Post* pilloried the now notorious area:

> In it, people live huddled together 281 to the acre. They live five and six in a single room that is part of some great slattern of a tenement, with seven or eight people in the room next door, and maybe eight or ten in the rooms above and below. The windows are often patched with cardboard. The stairs are narrow, dark at all times and befouled not only with mud and rain. Commonly there is one lavatory for thirty people and that with the door off.

The article continued:

> The Gorbals has no large industries and few small ones, where its residents may find work. There are always a lot of Gorbals residents on Public Assistance, either through sickness or in the interval between one short job and the next.

One local girl told reporters, 'I hate it in the Gorbals. If I meet anyone new, I have to give a false address.' Another girl said:

> We're eight in the one room. We go to bed in relays. My elder brothers walk around the back court while we girls undress. Then they come back and kip down on the mattresses on the floor beside us. The cat sleeps with us. If a rat runs over the blankets, he springs out and has it.

Later, in 1965, local minister Richard Holloway told reporters, 'I've married young couples who were obviously very much in love, and then seen that love wither simply because of the conditions they had to live in.'

However, most Gorbals people regarded appalling living conditions as a fact of life and just got on with it, saying it was far tougher in the old days. Some of them had been brought up in families as large as twelve in a single end: one room and kitchen. It was not uncommon even in the 1960s for six people to share such cramped accommodation.

Back home in Crown Street, my parents were usually out working at the weekends. My father had a job as a chef in a restaurant in the centre of Glasgow and my mother worked as a barmaid in Maryhill. This meant that most weekend nights me, my brother and my friends had the run of the house. But my mother warned me, 'When Ah go tae work and ye're in the hoose yirsel, never ever open the door tae anybody, no unless ye know who they are. If a knock dis come tae the door, always shout, "Who is it?" Ye can never be too careful, there's a lot of nutcases oot there.' I was only about nine at the time, and my brother two years younger. But we loved the privilege of having such freedom from parental control. We would all sit down in front of the black-and-white television absorbing the variety of programmes.

Some TV shows even featured Glasgow's favourite patter merchants, like Francie and Josie or Chic Murray. Francie and Josie were comic heroes to the Gorbals children. The comedians Rikki Fulton and Jack Milroy dressed up as teddy boys and turned oan ra patter. Their catchphrases, such as 'Hellorer, china', became everyday greetings between local people. By comparison, Chic Murray's patter was surreal. He looked like the older men on the streets of the Gorbals with his tartan bunnet and sardonic expression. When he was on TV, we couldn't stop laughing. He'd say things like: 'My girlfriend's a redhead – no hair just a red head'; 'She had been married so often she bought a drip-dry wedding dress'; 'I went to the butcher's to buy a leg of lamb. "Is it Scotch?" I asked. The butcher said in reply, "Are you going to talk to it or eat it?" I then asked him if he had any wild duck. "No," he responded, "but I could aggravate it for you."'

One day, I was on a bus going from the Gorbals to the city centre when I noticed Chic Murray chatting to the passengers, and he had them in fits of laughter. He was in full flow. An elderly couple asked him, 'Whit hiv ye been up tae, Chic?' and he replied, 'I went to the doctor and he told me I only had three minutes to live. I immediately asked him if there was anything he could do for me. He replied that he could boil me an egg.' He continued, 'I rang the bell of this small bed-and-breakfast place and the landlady appeared at an outside window. "What do you want?" she asked. "To stay here," I replied. "Well, stay there, then," she said, and shut the window.'

He looked as though he had been having a good day out and his performance on the Corporation bus was first class. When he got off, the passengers gave him a round of applause. As I watched Chic Murray walk along Argyle Street with a slight swagger, I thought that, as in the song, Glasgow really did belong to him. It was then I realised that Glaswegian is the lingua franca of Scottish comedy. Everybody in Scotland laps it up and it is perhaps for this reason that the biggest names in Scottish comedy have been from Glasgow.

By 1967, television technology was changing fast. My mother and I were taken aback when we saw the first colour televisions in a shop window. She was transfixed at the bright colourful images on the screens. She said, 'Ah cannae believe it. It's jist like being at the pictures. Who wid have thought we wid have seen such a thing in our lifetimes?' A neighbour was one of the first people in the Gorbals to have a colour television. All her relatives and neighbours piled in most nights to gather round the screen. She complained to me, 'It's funny, but when Ah had an auld black-and-white telly, naebody used tae knock on ma door. But noo Ah've got a colour telly, Ah've got a lot o' new-found friends. Ma man bought it wi' his redundancy money fae the shipyards, noo he cannae even get near the screen.'

Hardly anyone in the Gorbals paid for their TV licence. When there was a clamp-down on licence dodgers, people threw a sheet over the TV. If an inspector knocked on the door, they told him that they 'didnae hiv a telly'. But if the inspector then persevered and accused them of not telling the truth, they replied, 'It isnae working.' They'd show him in to the front room where the TV would be covered with a blanket or sheet. The TV inspectors surely got fed up with this ploy, as it must have seemed as if almost every television they came across in the Gorbals was covered with a sheet.

Late on Friday and Saturday nights, my mother arrived back from her barmaid's job. She regaled us with stories about what the characters in the pub had got up to. My father also sauntered in after a busy night at the restaurant. He usually had a pile of food that he had appropriated from the restaurant, including steaks and chicken. He wasn't in a well-paid job, so he and the other chefs just took it for granted that pilfering food was a way to supplement their meagre incomes. He often said, 'We might be living in the Gorbals, but we're livin' aff the fat o' the land. Besides, it disnae matter where ye live, it's how ye live that counts.' One night he said he had almost got the sack: 'Ah had jist finished work and wis standing at a bus stop and waitin' tae go hame when the restaurant manager stood beside me. Ah got a bit nervous and two steaks fell oot fae the inside o' ma coat ontae the pavement. Ah thought Ah faced the bullet there and then but the manager burst oot laughing and said, "Ah've heard of it raining cats and dugs but never sirloin steaks!"'

Some weekends, we escaped from the Gorbals by going to visit my grandparents in Househillwood. We walked the several blocks from Crown Street to the 48 bus stop in Eglinton Street. One dark winter's night, me, my brother Ross and my father were walking along Bedford Street to catch a bus when we noticed a drunk young fellow punching hell out of an old man. My father was holding our hands but when he saw what was happening, he said, 'I'm no standin' for that, he's takin' a liberty.' He let go of our hands and rushed over. He stuck the head on the young fellow and then knocked him out. I had never seen my father in action before and he was an impressive sight. Afterwards, the three of us ran off to the bus stop with police sirens blaring in the background. Luckily for us, a bus arrived within a few seconds.

On the bus, my father was serious at first when he said to the two of us, 'Boys, Ah had nae alternative but tae dae somethin' aboot it. Otherwise he would have killed the old guy and I don't like liberty-takers. But don't be tellin' yir granny whit happened, it would gie her a heart attack.' Then he gave out a mischievous laugh as the bus trundled away from the dark and wild streets of the Gorbals.

On that walk between Crown Street and Eglinton Street, there were always incidents happening. One cold winter's night, two big men approached me and my father. They were about the same age as my old man and were reeking of cheap wine and beer. I could tell that my father didn't like them and that the feeling was mutual. He had recognised them from his youthful gang days.

One of the men looked rather menacingly at the two of us before saying in a cheeky manner to my father, 'Hey, is this your boy? Did you no tell him that it was me who taught ye how tae go tae town?' In other words, it was an insult. He was saying that he'd taught my father how to fight. The expression on my dad's face was a picture. He turned red with anger. I could tell he wanted to beat this fellow up right then and there. He decided to take the matter a step further by staring him right in the eye and saying, 'You taught me how to go tae town? Aye, on the f****n' bus. Why? Dae ye want tae dae somethin' aboot it?'

It was then I realised he would have no difficulty in setting about these two chancers. All of a sudden, the two men laughed it off, saying they were only joking, and walked away. My father said to me, 'Those bampots are aboot as hard as a couple o' jelly babies. Ah'm like an elephant and Ah never forget, so we'll see whit happens when I meet them next time.'

A few weeks later, my father was in a Gorbals pub and bumped into the men again. This time, he was the one that was under the influence and the two fellows were sober. He gave one of them a doing and the frightened friend ran away. My father said to me, 'Whit a pair o' mugs. They thought they could pick on me because Ah wis wi ma wee boy. But last night Ah showed them how tae go tae toon . . . in an ambulance.'

Chapter 3

LIFE, DEATH AND LOVE

Living in the Gorbals could be an uphill battle. Because of the poor living conditions, health problems were rife. When I was about ten, there was a massive outbreak of head lice, with my family and hundreds of others – men, women and children – afflicted. We had to purchase from the chemist a fine-tooth comb to expel a significant quantity of lice from our hair. As I counted the number of creatures being ejected from my head, I thought, rather bizarrely, that they had turned it into their own personal jungle. Because of epidemics such as this, almost every other boy at school, and some girls, had a shaved head. Luckily, I eradicated the problem in time, so I never had to suffer such an indignity. To me, a shaved head signified true poverty. Children who had impetigo also had all their locks cut off. One of the boys at my school had his hair cut off and his head covered in purple lotion, and the pupils taunted him, shouting, 'Here comes Scabby Heid! Baldie! Baldie! Baldie!' For years after that he was nicknamed 'Scabby Heid'; in fact, most people forgot what his real name was. The treatment for scabies was just as humiliating. You were instructed to take all your clothes off and then had a sticky white solution plastered all over you with a large brush. One boy said to me in the playground after this treatment, 'They painted me aw over wi this brush, it wis as if they were decorating a room.'

TB was a disease that was talked about in tones of hushed dread. It had killed hundreds of inhabitants of the area over the years. Entire families were wiped out by TB and many people spent years in bed in a 'fever hospital' battling the condition. Some talked of experimental operations in hospitals for the illness. The patient had a lung filled with oxygen until it eventually collapsed, and was treated from there. Older people told me that years before, pre-NHS, many Gorbals people had gone to meet their maker simply because they did not have the money to shell out for a doctor. Some had been more far-sighted and paid threepence a week to a friendly society to guarantee medical care. They had to, as at that time the Gorbals was rampant with diseases like diphtheria, scarlet fever and rickets. In the bad old days, sick people with no money were experimented on with new-fangled drugs, some of which had beneficial effects, many of which hadn't. We considered ourselves quite lucky, because the older generation told stories that were alarming.

A number of them said they had had to have surgical treatment in the house because they could not meet the expense of going to hospital. An 80-year-old Gorbals pensioner said to me that when he was younger he had had one such operation for a broken leg. He wasn't exactly precise about the surgical procedure but said the quack doctor had used very primitive techniques, including pieces of wood and a bottle of whisky as an anaesthetic. 'After I was knocked oot by the whisky, he messed aboot wi ma leg. When Ah woke up wi a terrible hangover the next day, ma leg wis throbbin' but after a few days it wis well on the way tae mending. Remember, in those years hiring a doctor was too expensive, so that wis ma only alternative tae get ma leg right again.'

Because of the influence of Scottish Highland and Irish culture in the area, bereavement was not looked upon with any great morbidity; rather a person's death was seen as a time to celebrate their life. When an old person passed away, the news spread like wildfire throughout the community because everyone knew everybody else. When, for example, a granny had taken her last breath, a relative would announce, 'She's gone,' and crying and wailing would fill the house. There were few houses with telephones in those days, so family members would rush over to the nearest phone box to contact the doctor and run to the priest's house with the sad news. Before the funeral, there would be a wake, and the funeral would be followed by another party to celebrate the person's life. There are far worse ways to go.

A wake was always the big 'send-off party'. Everyone got drunk, had a sing-song and told humorous anecdotes about the deceased person. People wandered over to the open coffin to have a last wee blether. The corpse often had a glass of whisky and a piece of cake placed beside it. Sometimes the body was taken out of the coffin and sat in a chair to be part of the party. In Catholic homes, incense was used, giving the house a chapel-like aroma. Children were not kept away and were even encouraged to attend the wake. On the streets of the Gorbals, the football matches came to a standstill, children stopped playing and singing, and men took off their bunnets to show reverence to the funeral cortège as it passed by.

Getting married also had its idiosyncrasies. A few days before the wedding, there was the inevitable hen party. The bride-to-be went on a pub crawl around the numerous bars of the Gorbals, with her friends loudly beating tin cans as they walked. Hen nights were run on pure alcohol, a fuel in which the Gorbals was not undersupplied. The women got louder and rowdier as they went from street to street. The bride would be dressed outrageously in a mock bridal veil and sometimes even a pair of handcuffs, and she was encouraged to kiss as many men as possible. After kissing the bride, men had to make a contribution to the collection tin, which went towards the cost of the hen night or a wedding present. The hen party might proceed to the groom's house and stand outside banging on tin lids, shouting his name: 'Jimmy! Jimmy! We know ye're hame, we know ye're hame!' And then they would all sing:

Ah widnae get married if Ah wis you,
If Ah wis you, if Ah wis you,
Ah widnae get married if Ah wis you,
Ah'd rather stay wi ma mammy!

Growing up in such a close-knit community, we were often confronted with evidence of the fickleness of love. One night, a gang of 30 hen-night women took over the streets, shouting, singing and bawling. We followed them and watched the bride-to-be kiss more than 20 men in less than an hour. The bride, with her eyes rolling in her head, staggered with her friends up the street, shouting to the men, 'Gee's a wee kiss and cuddle.' Most men were only too willing to kiss a pretty young woman. But for one man, whom we recognised as an ex-boyfriend of hers, the kissing lasted far longer than expected. It became a real winchin' session on the street. The bride wandered off with the fellow and her drunken pals egged her on. The couple then disappeared up a dark lane. The next day, she didn't bother to turn up for her wedding. She ran off with the other fellow, ending up pregnant, much to the amusement of all the gossips.

Many women feared the worst fate in life – being left on the shelf and ending up as a faded spinster. One story went that a local spinster was attending the funeral of a friend's husband when another pal remarked, 'That's the fourth husband Aggie's had cremated.'

'Wid ye credit that!' exclaimed the old maid. 'Ah cannae get a man . . . an' she's burnin' them!'

For married men, the old tradition of bringing their pay packets back home 'unbroken' – still sealed – carried on. A man finished his week's work and then laid his pay packet on the kitchen table for his wife. She worked out the week's running costs – her housekeeping money – and handed her husband his pocket money. It was an excellent idea to follow this routine, because those who didn't usually ended up getting intoxicated in the pub. They might blow a whole week's wages there, leaving their unfortunate wives and children to struggle for another week.

Some men, however, developed their own ways to break the rules and double-cross their wives. They splashed out on empty pay packets from a local stationery shop and when they got paid on a Friday, they swapped their real wages for false ones in the new packet. But the women weren't that stupid and they got wise to the trick. Many a broom handle, coal shovel or chamber pot was put over a man's head when his wife discovered she had been conned. She'd shout, 'Ya dirty rotten cheatin' swine! Imagine daein' such a thing tae yir wife and weans!'

In Gorbals families, it wasn't uncommon for the woman to wear the trousers. There was many a hen-pecked man who was always living in fear of her indoors but the women did a magnificent job feeding and clothing their families. They tried to provide a decent home for them on a pittance. Hundreds of families lived daily trying to cope with unremitting hardship. But a number of women just gave up after years of struggling with poverty and their men pishing their wages up against the wall. Marriages split up because of a lack of money; as the old saying went, 'Love goes oot the windae when there's nae money comin' in.'

Because of all the dramas unfolding in the Gorbals, there was no shortage of gossip. The numerous female gossips knew everything that was going on. If you wanted to hear the latest news about what was happening up the next close or street, you simply asked someone like Wee Maggie. She congregated with her fellow scandal-mongers in the street every day and the conversation usually started off with: 'Oh, here, did ye hear aboot . . .' The recipients of the gossip gasped and shook their heads in disapproval. Wee Maggie was like a human version of the *News of the World* and always had a scandal brewing about somebody or another. Her husband was a hen-pecked fellow who chain-smoked all the time. He stood on the street corner and joked about his wife's antics saying, 'The fastest ways for the news to get aroon' the Gorbals are telegram, telephone, television . . . and tell ma Maggie!' Her speciality was local sex scandals. Other women blushed as she exposed the latest sordid story in a whisper. Her tales were usually about loose women with low sexual morals.

There was absolutely no sex education at all in the Gorbals but some girls tried to stay virgins until they got married. It was supposed to be the trendy swinging '60s but the Pill and condoms were shunned by most of the Catholic population. The Church forbade contraception and encouraged the rhythm method instead. It was no wonder there were so many children running around.

Unwanted pregnancies could lead to women having back-street abortions. We had heard of an ex-nurse who carried out terminations in a tenement in Florence Street but she had very antiquated methods. Rumour had it that she even used a long crochet hook as a medical instrument. Many young women were very naive about sex and giving birth. They did not have a clue what happened. There were stories of young women in labour saying to the nurse, 'Cut ma belly open,' and the nurse having to explain, 'It disnae happen that way, hen – it comes oot where it came in.'

A woman who opted to have a sizeable family was considered to be either a 'good Catholic' or a 'lazy Protestant'. After a couple got married, the wife was expected to fall pregnant within six months at most. Friends and neighbours would stop her in the street, asking, 'Are ye expecting yet?' If the newlywed said, 'Aye!' the reply was, 'Oh, that's smashin', Ah'm really made up fur ye.' In the past, Gorbals babies had usually been born at home in the tenements, but by the 1960s, women were encouraged to have their babies in hospital, as it saved doctors and midwives traipsing up and down the cold, dark tenement stairs.

Dr George Gladstone Robertson spent 47 years in the area and published a memoir of his time there, entitled *Gorbals Doctor*. In the book, he recalled delivering a woman her 22nd child. In all, the woman had had 26 pregnancies and 4 miscarriages. He also wrote of women regularly appearing at his surgery covered in bruises caused by drunken husbands.

A fellow who was going to prison for a while would purposely make his wife pregnant. It was a form of insurance – he knew she wouldn't be messing about with another man while he was behind bars and would be there for him when he got out. Men in general had a macho attitude towards pregnancy and childbirth. As a woman lay struggling to give birth, her man would be in the pub with his pals getting drunk. The first he would hear about the outcome of the birth was from a relative or neighbour, who'd accost him after the pub with the good news. No man I knew ever attended the birth. Now it's commonplace, but it just wasn't the done thing in the '60s. Also, because of the macho culture, I never saw a man pushing a pram. It was considered the man's job to go out and earn the money and the woman's to stay at home, make the dinner, wash the dishes and look after the weans. One day, I heard a young married guy from Thistle Street say to his pals, 'Ah wis up the toon the day and Ah couldnae believe it when Ah saw this big bampot pushin' a pram. Whit is the world coming tae? Diddies like that must be no right in the heid. It's a woman's job tae push a pram, always has been, always will be. They'll be wantin' us pushin' their prams and bein' at the birth next. No way is that ever gonnae happen wi me!'

Chapter 4

PARTY TIME

When I was a boy, my parents had the occasional party at the house, inviting neighbours, friends and relatives in for a good bevy, especially if both of them were working and had full pay packets. People sat around the living room, which we called 'the big room', exchanging patter and regaling all and sundry with their anecdotes.

My grandfather, my father's father, an old Glasgow seaman, told a variety of outlandish tales which had people laughing. 'A toff walks intae a shop wi a big hat on and orders some pipe tobacco. He goes ootside and tries tae light his pipe but it's too windy. So windy, in fact, it blows his hat aff. A big dug runs oot the shop and, seein' the toff's hat blowin' along the ground, he runs after it and takes a big bite oot o' it. The toff goes back intae the shop and says tae the owner. "Your dog has just taken a bite out of my expensive hat and I want compensation." To which the shopkeeper replies, "It's got nothin' tae dae wi me if ma dug takes a bite oot yir hat. It wis an act o' nature." The toff says, "I don't like your attitude." To which the shopkeeper replies, "It wisnae ma hat he chewed, it wis your hat he chewed!"'

After the storytelling, everyone would have a sing-song. My grandfather would shout, 'Order, order, wan singer, wan song,' and it would begin. Favourites included 'I Belong to Glasgow', 'It's a Sin to Tell a Lie', 'Among my Souvenirs' and Al Jolson's 'Mammy' or 'April Showers'. (Al Jolson tended to dominate sing-songs when I was a boy. This was due to the impact that the 1947 movie *The Jolson Story* and its sequel *Jolson Sings Again* had had on Glasgow audiences.) One night, an old Gorbals widow called Bessie brought the whole party close to tears with her version of 'Among my Souvenirs', sung in a voice that reflected a life of hardship and sadness. Bessie told the party, 'Once Ah had a man who loved me and a wee hoose where Ah felt loved. But when ma man passed away, Ah wis left wi nothing except fur ma memories o' aw oor good times thegether.' Still wearing her headscarf and coat, she began crying as she sang. Everyone at the party agreed that this woman could bring tears to a glass eye.

There was sometimes a punch-up at these shindigs, but they were few and far between. Usually everyone ended up being mellow and sentimental, with the cheap wine flowing. The big social gathering of the year was Hogmanay, to see the old year out and the new one in. Everybody in the Gorbals, no matter how young or old, was geared up for the bells. As late as the 1960s, in Scotland New Year's Eve was still rated a bigger event than Christmas, which the English celebrated far more than us. For hundreds of families in the Gorbals, New Year's Eve morning began with everybody joining in on a major cleaning exercise. The close stairs were given a good scrub to give first footers a clean and tidy welcome for the New Year. The preparations went on all day and finished as late as an hour before the midnight bells.

Like the rest of the Scots, Gorbals people were superstitious. It was believed – and still is in many households in Scotland – that it was bad luck to have a dirty house at midnight. My mother often spent the 31st making sure the flat was spotless. Tall, dark first footers bearing a lump of coal and a carry-out were welcomed with open arms because this meant the house and family would be prosperous in the new year. The last thing any house wanted was someone with red hair turning up as the first footer, because that meant bad luck for the coming year.

Before midnight, there was usually a large steak pie heating slowly in the oven, with pots of mashed potato, peas and other vegetables. There was also a pot of Scotch broth simmering slowly, made with a fresh ham hock. Part of the tradition of Hogmanay was having the television on and watching *The White Heather Club*. It represented Scotland in a completely stereotypical fashion. It had men and women dressed in kilts reeling about with the sound of fiddles and bagpipes in the background. *The White Heather Club* was considered good clean fun at the New Year but it was unlike our reality. There were no drunks, fighting, drinking or swearing on the show, and Andy Stewart, the host, always did his funny turn every year, singing 'Donald, Where's yir Troosers?', complete with Elvis impression. My uncle Mick commented to me one Hogmanay, 'Aye, Andy's a rare wee singer but he's livin' in a fantasy world. Ye never see anybody wi a scar or a carry-oot in *The White Heather Club*.'

One particular New Year's Eve ended up turning into a debacle. My father had sent Uncle Mick to the licensed grocers to get some wine for the party. Mick duly came back with 15 big bottles of El Dorado and a Mother's Pride loaf. On seeing this, my father jokingly asked, 'Hey, Mick, why did ye buy aw that bread?' As a result of the purchase, my parents had had too much wine far too early when a woman from downstairs called in to join the celebrations. She was known for having a mouth bigger than Grangemouth and she had a voice that could be heard a mile away at Glasgow Cross. For all her faults, I could see nothing really wrong with her. She was a bit loud and mouthy but friendly enough to me. However, I could tell my father did not like this woman. He had barely spoken to her during the two years she had lived in our tenement.

As the night progressed, everyone began dancing to Engelbert Humperdinck's 'Please Release Me'. By this time, the mouthy woman had had far too much to drink and was at loggerheads with a few people, saying things like, 'Ye're talkin' through a hole in yir arse.' My father was fed up with her obnoxious patter and threw her out. She began shouting, 'Ye've no heard the last o' this! Ah'm gonnae get ma man and his brothers tae sort ye oot.' There were a few minutes' silence and then, as if nothing had happened, everyone began laughing and continued with a sing-song.

But after more wine had flowed, my father and uncle became obsessed with the woman's threat and went down the tenement stairs to her house. I followed them as they began banging loudly on her door but there was no reply. She was obviously too frightened to open her own front door, knowing there were two wine-fuelled maniacs outside. They then headed down the close to the back court, picked up a couple of bricks and threw them through the woman's window. I thought this was madness, as she lived just below us. The smashing of the glass made an incredible noise; it was a frightening sight for someone as sober and as young as I was – I had just turned 12. Moments later, we all went back upstairs to the celebrations.

Two policemen arrived at our house later that New Year's Day, well after the party had ended. My father and mother both had terrible hangovers when they were charged with breach of the peace. But the matter did not end there. A week later, my parents were out working on a Friday night and an uncle had been sent to look after us 'jist in case there are any comebacks'. Around eight o'clock, there was a knock at the door. We did not open up but a man outside was saying he had found a comic in the close and asking if it belonged to me or my brother. More suspiciously, he asked us to open the door. I immediately sensed something was up and shouted for my uncle. The next minute, there were four men outside trying to kick the door in. My uncle pulled out an air pistol and began firing it at the men through a little window near the toilet. As they were shot, the men screamed out and then ran off. My uncle said they were relatives of the mouthy woman who had come to seek their vengeance. Not long afterwards, the woman moved away from the Gorbals; she must have feared further repercussions. I don't blame her. She was right: there would have been serious repercussions.

My father appeared at Glasgow's Central Magistrates' Court, where he was fined £25 for breach of the peace. The magistrate warned him about future behaviour. 'There are far too many incidents like this happening in the Gorbals at the New Year,' he said. 'This sort of behaviour cannot and will not be tolerated. Why can't the people of the Gorbals just behave themselves for a change instead of shouting and bawling, fighting and putting bricks through their neighbours' windows? It's time you all grew up and acted with a sense of common decency towards each other. Let this fine be a lesson to you.'

As we were walking away from the court, my father, who had looked sober and solemn in a suit, shirt and tie during the course of the proceedings, gave a big wide smile and began singing, 'Please release me, let me go!' We began to laugh at the absurdity of the situation, at the madness of living in the Gorbals.

Chapter 5

THE STEAMIE

When my mother had to do a washing, she, like hundreds of other Gorbals housewives, made her weekly trip to the steamie, or the bagwash, as some locals referred to it. The steamie was an integral part of Gorbals life. Even by the mid-1960s, few people in the tenements had their own washing machine. So the Glasgow Corporation-run steam-driven washhouses were the best option. At the turn of the century, local people had been used to doing their washing in the back courts and hanging their laundry on lines. But by the '60s, the public washhouses were booming because they made hard work easier for the Gorbals housewives. They went to the steamie wearing their coloured printed headscarves with prams loaded with washing, entering a world that was completely alien to the male species. The men had their pubs and football, and the women had their steamie and bingo. For many women, it was a day out and a chance to have a good blether with their friends. The scene inside looked very Victorian, with rising clouds of steam and hundreds of bare arms rising and falling while doing the washing. Posh ladies, of course, never went to such places; it was a strictly working-class domain. Sometimes, when I waited for my mother at the steamie in Rutherglen Road, I was overwhelmed by the images and banter. The patter was usually up to the best standards of any music-hall routine, with gossip and scandal flying everywhere.

'Hey, did ye hear aboot her up the road wi aw the carry-oan?' one woman might say to another.

'No, whit happened? Is she still up tae her auld tricks, is she?' would come the reply.

'Aye, well, ye know that fella she left her man fur? Well, he's ended up in jail again, leavin' her withoot a penny tae her name. It's the weans Ah feel sorry fur, they look like they're neither washed nor left alane. She should never hiv left her man. At least he hid a decent job, no like that big chancer she got involved wi. He's a right waste o' time.'

'Aye, she must hiv been the biggest mug in Glesga tae get involved wi the likes o' that. Ah widnae even gie him a nod in the desert, the big no-user that he is. When he comes oot o' jail, somebody should gie him a doin' fur whit he's put that woman and her weans through. A doin' might knock some sense intae him.'

'Aye, ye're right, he's got it comin' tae him but she wis stupid getting involved in the first place.'

'Mind you, it takes two tae tango.'

'Aye, two tae tango, an' that waste o' space his led her a right merry dance.'

The banter would continue on the subject of another woman and her partner.

'Hiv ye seen her man? A wee stupit-lookin' bachle, he is. He's got a gub like Fort Knox.'

'Whit dae ye mean, Isa?'

'He's got that much gold teeth in his mooth he's got tae sleep wi his heid in a safe.'

'Ah don't know whit she sees in that wee bachle. He's no the size of a tuppence.'

'Ah bet he wis aw she could get.'

'Aye, she wis that long on the shelf she had coarns on her arse.'

There would be silence for a few seconds as they handled their washing, then one of them would say, 'Did ye no hear aboot Wee Maggie gaun' tae hospital wi her bad back? Poor wee soul, the doctor told her . . .'

The steamie was divided into stalls, each with its own boiler, sink and iron. They were let out for as little as a shilling an hour. When the washing was done, there was a drying rack called a horse. It was a long contraption and you pulled it out, put the clothes on it and then pushed it back in to what was a quick-drying chamber. The steamie was a forerunner of the modern launderette but today's launderettes could never match the social interaction that occurred between neighbours and friends there. They took great pride in their washing. It became a sign of self-respect to have an immaculate load in your pram as you headed home. A huge insult in school was for someone to shout, 'Your maw's so clatty, her drawers are banned fae the steamie!'

If any mother didn't have the time or inclination to go to the steamie, she usually washed the clothes by hand in the kitchen sink. They were then left to dry on a mechanical pulley in the hallway. The pulley was raised and lowered by ropes and the clothes hung on its wooden frame. One of my pals had been reading too many Superman comics. He was convinced that if he put himself on the pulley and got all of his pals to pull it up in the air, he would have the same feeling as Superman and be able to fly about the house. He was only up in the air a couple of seconds when the weight of his body cracked the pulley and he went flying to the ground. There was a great palaver afterwards when his mother found out.

Once the day's work was over, the Gorbals women often got together en masse to go to the bingo. Watching hundreds of women rushing off to the bingo on a cold winter's night was a sight to behold. They generally headed to the Palace Bingo Hall in Gorbals Street. They got the nickname 'the Bingo Bellas' because we often heard a woman shout to her pal, 'Hey, Bella, ur ye gaun tae the bingo?' On busy nights, there were hundreds of them queuing outside of the Palace. They came from all over Glasgow and some nights the queue stretched right up Gorbals Street. Most of the Bingo Bellas had hard lives, coping with an existence of poverty, but through it all they had a magnificent spirit of optimism, hoping that they would win 'the big snowball', the jackpot, which could run to hundreds or even thousands of pounds. A win helped to transform their lives for a few weeks at least. Alex had a good joke about the Bingo Bellas: 'How dae ye get yir granny tae shout "bastard"? Get another woman tae shout "house" before her.'

We usually had good banter with the Bingo Bellas, wishing them luck and shouting, 'Hey, missus, we can tell a winner and you're gonnae be one tonight!' Their faces lit up, because many of them were only too glad of the encouragement. It was a psychological ploy. I believed that if any of them won the big snowball they would treat me and the rest of the boys as lucky charms and give us a bung from their winnings. Much to our surprise, this happened often, with the winner approaching us shouting, 'You brought me luck tonight, son.' She would then press money into our hands. When the bingo players won money, they knew how to enjoy themselves. A crowd of them headed to the nearest pub at Gorbals Cross for a celebratory drink. At the start of the night, a Bingo Bella arrived on a Corporation bus but if she was a winner, she took a taxi back home, feeling like a queen, victorious with her snowball. For many of the women, going to the bingo really was the light at the end of a very dark tunnel.

Chapter 6

FOOTBALL CRAZY

The Gorbals and the rest of Glasgow was, and still is, 'fitba crazy, fitba mad'. Celtic and Rangers dominated everyday conversation; football was the passion of most men. From an early age, almost every boy in the Gorbals dreamed of playing for the Old Firm. Parks like the Glasgow Green and Govan Park had up to 50 pitches each. Every one was jam-packed with young footballers on Saturday and Sunday mornings. When I played outside-right for the school's Under-14s, more than 100 Gorbals people turned up to cheer us on. Whole families would be there to give their support. Indeed, one Scottish Junior Cup match attracted an astonishing 10,000 people. During the summer, the back courts and streets were full of youths playing football. The games carried on for hours on end and even all day. I remember playing football one day with a bunch of other kids for twelve hours, nine in the morning till nine at night. We used our jackets or jumpers as goal posts. Each match usually consisted of ten goals half-time and twenty-one the game. Some of the boys were so good they went on to become professional, as football scouts were always scouring the Gorbals for talent; top footballers like Tommy Docherty and Pat Crerand came from the area.

Wee Peter, one of the local lads, was addicted to football and played it night and day in the spare ground near the Brit in Thistle Street. He was a good player and was tipped to go far, because many believed he had the cleverness and enthusiasm to turn the sport into a full-time job eventually. If anyone was destined to play for Celtic, which was his aspiration, it was probably Peter. But one day, he was in a dangerous tackle with a big gawky fellow and broke his leg rather badly. The big guy was a hopeless football player and was better at kicking people than the ball. We were all disgusted at what had happened to Peter, and the gawky guy kept out of our way because he feared repercussions. Peter had to have a series of operations on his bad leg and was on crutches for more than a year while doctors tried to sort it out. But his hotheaded nature ensured his injury did not put him off playing football. He still took part in the back-court football matches on his crutches, sometimes in goal and even as an attacker, and kicked the ball with his one good leg. Word got out and spectators turned up solely to watch Peter playing on his crutches. It really was an extraordinary sight. The story about Peter still scoring goals on his crutches attracted much media attention and he appeared in the papers. After seeing a photo of Peter, crutches and all, Celtic manager Jock Stein told the press that he was so impressed with the lad's determination he would give him a trial when his leg eventually healed. Of course, his injury meant that he would never become a professional footballer but encouragement like that gave Peter the confidence to carry on and made him a bit of a celebrity in the Gorbals. He told me, 'Ah can still play wi one leg far better than some people can play wi two.' I just thought if he wasn't a classic example of football crazy, football mad, then who was?

All the boys lived in football fantasy worlds where they tried to emulate the skills of their heroes, such as Celtic's Jimmy Johnstone or Rangers' Jim Baxter. When Celtic played Rangers or Scotland played England at Hampden Park, you could always hear this terrific cheering coming from the stadium, just a few miles up the road in Mount Florida. This noise was nicknamed 'the Hampden Roar'. When Celtic played Leeds at Hampden in 1970 in the European Cup (with a British-record crowd of more than 130,000 people), it sounded like everybody in Glasgow was roaring at the one time. I have never heard a hullabaloo like it since.

On 25 May 1967, the streets of the Gorbals looked like a ghost town as people remained in their houses or crammed into pubs to watch the European Cup final. During the 1966–67 season, Jock Stein had turned the club, which for years had been living in the shadow of Rangers, into the best team in Scotland. They were the first British side ever to reach the final of the European Cup. Inter Milan were the favourites to win the competition and another side might have been intimidated by the formidable Italians, but on that May night, Celtic went into the match as a team of gallus Glaswegians, having won the League, the Scottish Cup and the League Cup. Now they were prepared for the biggest prize of all. Hundreds of green-clad supporters left in coaches from the Gorbals for the trip to Lisbon in Portugal, where they joined an army of thousands of Celtic fans.

All the pubs were bursting at the seams with people egging on Celtic as the match was broadcast live on TV. I sat in front of our big black-and-white telly and watched the drama unfold. There was a major setback when Inter scored from a penalty only minutes into the game. We had to wait till the second half for the equaliser. Tommy Gemmell fired home from twenty-five yards to level the score. With six minutes to go, the drama came to its culmination as Bobby Murdoch's shot was deflected past the Inter goalie by Steve Chalmers. It was the winner – and Celtic's 200th goal of the season. The mobs in the pubs and streets of the Gorbals roared at this tremendous victory. Celtic had become the first British team to lift the most prestigious trophy in European football.

When the game ended, I took a quick look out of the window and the very first thing I saw, to my astonishment, was a man holding his Celtic scarf behind his head and kissing the pavement in Crown Street. It was as if the unsanitary streets of the Gorbals had suddenly become hallowed ground. This was definitely football crazy, football mad. Everyone came out of their houses and the pubs, and suddenly there was a gigantic social gathering. Thousands of people marched up and down the streets waving scarves and flags and joined in on a rousing sing-song. The procession began to sing:

> Hail, hail the Celts are here,
> What the hell do we care now?
> For it's a grand old team to play for,
> For it's a grand old team to see.
> And if ye know the history,
> It's enough to make your heart go oh, oh, oh.
> We don't care what the animals say,
> What the hell do we care?

> For we only know that there's gonna be a show,
> And the Glasgow Celtic will be there!

Then they started singing, to the tune of Al Jolson's 'Mammy':

> Celtic, Celtic,
> I'd walk a million miles
> For one of your goals,
> Oh . . . Celtic!

There was a carnival atmosphere. I saw men dancing jigs on the top of a pub roof in Florence Street, shouting: 'Oh, ya beauty, ye! We arra champions o' Europe and naebody can beat us, no even the Italians.' The *Daily Record* van arrived at around ten o'clock in Cumberland Street, with the first editions proclaiming Celtic to be European champions. The jubilant mob hijacked the van and thousands of free *Daily Records* were thrown into the air. Waving their free newspapers, the crowd sang 'You'll Never Walk Alone' like a gigantic Gorbals choir.

A few days later the supporters arrived back in town with their outlandish tales and mementos. Many of them looked as though they had been drinking non-stop since the final whistle. I had never seen so many drunk and hungover men at the same time. At school, boys proudly arrived with huge clumps of pitch grass that their fathers or uncles had dug up after the game. The grass was passed around and some pupils kissed it to celebrate the victory over the Italians. The ironic thing was that in the Gorbals there were numerous Italians selling ice cream and fish and chips. I had the feeling, particularly on that night, that I was living in one big international village, where we were all truly European.

The Gorbals was often swathed in green, due to the Celtic and Irish influence. But my father banned green from the house, saying it was an unlucky colour. One night, there was a green Celtic scarf, left by one of my friends on a chair near the coal fire. My father had just come through the door and my mother was making tea. Then a spark flew from the coal fire onto the chair and set the scarf alight. My father rushed to the sink and put the flames out with a saucepan of water.

He shouted at me, 'Whit did Ah tell ye? Don't hiv that colour in this hoose ever again. Green has never been a lucky colour for me.' The words took me aback, because everywhere I went in the Gorbals there was always the colour green in the house. One night, after a few glasses of wine that had made him mellow, he explained why he hated the colour so much: 'When Ah wis wee, every time Ah wore green, somethin' bad wid happen tae me. I knew o' one guy in Partick who wis doin' really well until he bought a green vase. Then, suddenly, his wife left him takin' the weans wi her. His health suffered and his business went doon the drain. But as soon as he threw the green vase away, his luck came back and everything and everybody went back to normal.'

He then pointed to a scar circling the rim of his nose and said, 'That is the main reason why Ah hate green. When Ah left school and got ma first job, I bought myself a green suit. I was walkin' alang the road wi ma flash green suit on, looking the business and feeling very much like wan o' they big film stars, when I saw two guys comin' towards me. I hid a fight wi wan o' the guys a few weeks before. They stopped me in ma tracks but I wisnae afraid o' them and was ready to go ahead. Just as I was about to get stuck in, one of them pulled oot an open razor and slashed me across the top o' my nose.

'Part o' ma nose fell aff onto the pavement, the two of them ran aff and ma face was covered in blood. Jist by chance, a woman was passin' by and she managed to put ma nose into a hanky. Ah was taken to hospital and, luckily, they managed to sew it back on again. But Ah put it aw doon tae wearing that green suit.'

Chapter 7

HUDGIES, BOGEYS AND MIDGIE RAKING

When I was twelve years old, a few months after my dad's appearance before the magistrate for breach of the peace, I had to attend the same court on a different charge. I had been standing on a street corner when me and my pal Chris saw a bin wagon stop. We decided to go for a hudgie, which hundreds of children did all over Glasgow. We jumped on its steps and gripped tightly onto the handles. The next minute, we were whizzing up and down the road, shouting and laughing hysterically. But when we were going down Gorbals Street, two policemen spotted us and began a chase. But they couldn't catch up and we poked fun at them as the big wagon sped off. A few hours later, I was standing in Hospital Street when the two policemen approached me and Chris and said, to him first, 'Did I no see you earlier on the back of a bin wagon?' Chris replied, 'Nah, it must have been somebody else, officer, Ah wis in the hoose wi ma mammy.' I thought I would show more bravado and said, 'Aye, it wis me. Whit's wrang wi hivin' a hudgie? There's nothin' fur us tae dae in the Gorbals anyway.'

It was the first time I had told the police the truth, breaking the unwritten Gorbals rule. Result? A summons to appear in court. I received a two-pound fine, with the magistrate threatening me with approved school and warning my parents about the dangers of young boys catching hudgies. 'Look, young man,' he said to me, 'you can't go all over Glasgow using bin wagons as a sort of taxi service. It's just not right and it's a matter of public safety. One of these days some poor boy is going to get seriously injured because of this stupidity. This time I will fine you two pounds but if you dare appear before me again, the sentence will be much harsher.'

The two pounds seemed a lot of money. I just thought that if crime didn't pay, neither did being honest with the police. My mother and father were mortified. It wasn't only because of what they considered to be a harsh fine but also because the magistrate who fined me was the same fellow who had fined my father a few months before. My mother shook her head on leaving the court and said: 'He must think this family are aw hawf mad. Mind you, he must think that anybody who decides tae stay in the Gorbals is no the full shilling anyway.'

My being fined did not stop me and my friends catching hudgies all over the place. We continued to ride on the backs of bin wagons, lorries and even ice-cream vans. We ended up all over Glasgow, in such far-flung locations as Govan, Partick and Kelvinside. Catching hudgies was like having our very own chauffeur-driven vehicles. It was also a good way to have a day out from the Gorbals free of charge. We weren't really frightened of the danger that was involved; although we had been warned on numerous occasions, we never knew of anyone being seriously injured that way.

One day, though, I made a huge mistake. A Mr Whippy-type ice-cream van pulled up in Thistle Street. As the kids queued up for their ice-cream pokey hats in the summer sunshine, I noticed a small bit of metal on the back of the van. I believed I could hang on to it for a hudgie. Some of my pals dared me to have a go. So I grabbed the little bit of metal and hung on. Then the van sped off far faster than I had anticipated. Luckily, there was no traffic behind me as I went flying through the air. The force of the ride sent me into a somersault, landing bang in the middle of the road. I bashed my face badly and there was blood gushing from my head and nose. I had bruises all over my body and it looked as though I had been beaten up.

Covered in blood and bruises, I staggered back to the house like an injured war veteran. My mother was shocked at my appearance. She shouted, 'My God, whit's happened tae ye, ye're in a helluva state, quick somebody call an ambulance!' My father began shouting, 'Aw no, whit hiv ye been up tae noo?' We all sped off to the hospital in the ambulance, leaving my brother to be looked after by the next-door neighbour.

The drama unfolded. The nurses on duty at the hospital asked me, 'Who hit ye, son, wis it yir mammy or daddy?' I said, 'No, they didnae hit me, Ah wis oot playin' and Ah fell aff a dyke.' But then a doctor turned up in his white coat and stethoscope. He said, 'This boy's injuries are not consistent with falling off a dyke. I suggest the police should be called in to deal with the matter.'

Two detectives duly arrived and they were convinced my parents had given me a good kicking. As I was from the Gorbals, they were sure I was a victim of child cruelty. They used the 'Mr Nice and Mr Nasty' technique on me. As I lay in the hospital bed, one detective, a thin, grey-looking man, gave me some chocolate and was very friendly. He said, 'Come on, son, tell us whit really happened. Who gave ye the doing? If ye tell us the truth, then we'll no be back tae bother ye any more.' I kept up my fabricated story that I had fallen off a dyke.

Next came Mr Nasty, an overweight fellow with extremely bad breath. 'Look, ya wee liar, tell us the truth. Who hit ye, yir auld man? Tell us the real story or we'll have tae dae ye for wasting police time.'

The detectives also questioned my mother and father for over an hour and they protested their innocence. I had no alternative but to admit to the police eventually that my parents were not to blame, rather it was a hudgie on an ice-cream van that had caused my sorry state. Later, my mother ranted and raved at me: 'Dae ye want tae see me and yir faither locked up in jail, framed fur somethin' we didnae dae, and you and yir wee brother in a home jist because ye want tae gallivant aboot Glesga on wagons?' I gave my word I'd give up hudgies; it wasn't worth the near-death injuries or the consequential trouble.

Around the time we were catching hudgies on the backs of ice-cream vans, there was an English guy we called Wee Fred going round the Gorbals and Castlemilk in one of them. He was a small but thick-set man with thick, dark curly hair and a broad Gloucester accent. He seemed to be quite friendly with all the kids, especially the young lassies. But he lost his rag one day when Chris tried to grab a hudgie from the back of his van in Thistle Street. Waving his fists wildly, he shouted to Chris, 'F*** off from there or I'll kill you!'

From then on, we surmised that there was something not quite right about him – not the full bob. Me and the boys even had a suspicion that he might be a child molester who used his ice-cream van to get in contact with young lassies. He was always boasting about his exploits with women, even claiming that his girlfriend was the Glasgow pop singer Lulu. 'I like the Barrowland, plenty of birds there and usually the Scottish fellas are too drunk to chat them up. That's when I make my move,' said Fred to us one afternoon, and he gave us a wink. But it was the kind of wink that did not amuse us. Being Gorbals street boys, we could tell most of his stories were a tissue of lies and that he lived in some kind of fantasy world. He said he had come to Glasgow all the way from Gloucestershire, as he had family and friends in the area. One of the older boys said to us, 'Dae ye see the way that wee English bastard is always chatting up the young birds? I don't think he's aw there. He'd better start watching his step, otherwise he'll get done in by wan o' their brothers or faithers. There's somethin' aboot that guy that gie's me the creeps. It's time that bampot got the f*** back tae England where he belangs.'

Wee Fred the friendly ice-cream man appeared in a story in the *Evening Times* a few weeks later. He had run over a boy with his van in Castlemilk, with fatal consequences, claiming it had been an accident. Not long afterwards, another ice-cream van driver told us Fred had gone back to his home in England. We never saw him again, but years later I instantly recognised Wee Fred when his picture appeared in the national newspapers and on TV. It was Fred West, the mass murderer.

If they didn't fancy the hudgie game, children in the Gorbals had their own handmade transport vehicles: bogeys. Many of the kids should have become successful engineers if bogey building was anything to go by. Bogeys were made from all sorts of materials, including old prams and rollerskates. A bogey was a fast, inexpensive and effective means to get around. The favourite design, if you couldn't get hold of a half-decent old pram, consisted of planks of wood nailed together, with small wheels or rollerskates attached. It had a sort of steering wheel at the front – a small piece of wood tied to a rope – and an old orange box for a seat. On our bogeys, we travelled all over Glasgow, getting involved in adventures. If we saw another bunch of youths with better bogeys than ours, and if they were not as tough as us, we became 'bogey pirates' and hijacked them. We made off in our new vehicles leaving our old bogeys with them as a consolation prize.

We also used our bogeys as getaway vehicles, transporting stolen goods under the noses of the police. We once skipped into a picture hall in the city centre and found an unlocked storeroom full of confectionery boxes. We made off with the haul, stashed in our bogeys, all the way back to the Gorbals, where it was shared out with our friends. My pal Alex said triumphantly: 'It wis smashin' glidin' through the toon wi panda cars and Black Marias passin' by knowin' we had a load o' knocked-aff chocolates and sweeties oan us. I mean, hiv ye ever heard o' anybody being arrested on a bogey? It widnae even cross the polis's minds tae search a bogey. They're the best getaway vehicles ever made.'

Alex regularly used his bogey to get up to all sorts of antics. He went to the city centre and crept in the staff entrances at the backs of restaurants and stole what he could. He often returned with wallets, purses, coats and jackets. Once he arrived back in the Gorbals with a stuffed deer's head complete with antlers. A passing drunk man bought it for two bob for his wife's birthday. Eventually, Alex's luck ran out. He had crept in the back of a restaurant and was stealing a purse from a coat but he was caught by a crowd of hysterical staff whom he had robbed from previously. He later got sent to approved school, for that and other misdemeanours, but what really annoyed him most was that the police confiscated his bogey as evidence and he never saw it again.

Every Saturday morning, from when I was about ten, Chris and I got our pocket money, usually about half a crown each. Chris was a wild, red-haired boy from a hard-working Irish family. He was always clowning around and always did things with a great sense of humour. Most weekends, I went to his house in the next block in Crown Street and we embarked on our adventures. Chris invented a side-splitting Saturday-morning ritual. He took a ten-bob note from his mother's purse (sometimes without her knowing); then he went to the first-storey window on the landing and said in a foreign accent, 'Money means nothing to me,' before throwing the note out of the window. We ran down the tenement stairs as fast as we could and tried to catch the ten-bob note before it hit the ground. Next, we headed off to the nearest baker's and bought two giant apple pies, which we devoured before going to the nearby flea pit, the George Cinema. Sometimes we also bought bags of cheap broken biscuits from a cut-price store called Lennox's before we went in.

As they stood in line for the pictures, often in the wind and rain, younger Gorbals children sang:

> Skinny malinkey long legs,
> Big banana feet,
> Went tae the pictures,
> Couldnae find a seat.
> When the picture startit,
> Skinny malinkey fartit,
> Skinny malinkey long legs,
> Big banana feet!

There were plenty of picture halls to choose from, including the George and the Bedford, or the Coliseum (with a gigantic widescreen) in Eglinton Street. The ABC Cinema chain also ran a Minors Club, charging sixpence for admission, offering a morning of escapism, with bold colourful posters declaring 'Films, Fun and Good Fellowship'. The ABC regulars had their own weekly song:

> We are the boys and girls well known as minors of the ABC,
> And every Saturday we line up,
> To sing the songs we love and shout aloud with glee.
> We love to laugh and have a sing-song,
> Just a happy crowd are we,
> We're all pals together,
> We're minors at the ABC.

Every Saturday the Gorbals kids queued up in their hundreds to see the latest adventures of heroes like Zorro or Dan Dare. These serial flicks inevitably ended up with the hero being left hanging on to a cliff or in some other death-defying situation. The excited young audiences had to wait another week and pay another tanner to see what happened to their hero.

We sometimes bought frozen-solid orange jubilees, which we sucked dry by the end of the afternoon's presentation. The jubilees often caused a major predicament, as they were dangerous missiles, frequently hurled at the screen, cinema staff or other cinema-goers. One afternoon, what appeared to be a full-scale riot broke out in the George Cinema when an usherette and the manager told the noisy children to shut up and quieten down. They were hit with a shower of jubilee missiles. The usherette ended up with two black eyes and the manager was covered all over in lumps and bruises. It was extraordinary to think that they could have been stoned to death by a throng of screaming children.

During the course of the picture, fights often broke out between the boys and girls. One Saturday, I was sucking a jubilee and watching the film when a rough-looking lad behind me put his feet up against my shoulders. I turned round to him and shouted, 'Get yir manky feet aff me!' He was smaller than me but I had underestimated him. He shouted, 'Who dae ye think ye ur? Ah can put ma feet where Ah want. It's ma picture hoose.' He then walloped me full on the nose, with blood going all over the seats. Blinded by the blood, I was helped to the toilets by an usherette, where it took half an hour to stop the flow with paper towels. I was lucky he had not broken my nose. I learned from then on never to go to the pictures alone. It was far better to be well-handed, with some of my friends, just in case any trouble broke out.

After leaving the pictures one Saturday, we got back to Chris's house. His father, a big Irishman who worked on the building sites, had found out about Chris's ten-bob scam. He was furious. He picked up a large sweeping brush and chased him all over the house with it. He was shouting, 'Ah'll teach ye! Ah'll teach ye no tae take money fae yir mother's purse.' Eventually, Chris had nowhere to run or hide except under a big double bed and even then his father tried to sweep him out. But Chris stayed under the bed for hours until his father just gave up in frustration. 'Whit am Ah gonnae dae wi that daft boy?' he said, shaking his head in frustration.

When Chris and I had some pocket money, we also went to the local sweetie shop and got a quarter of what we fancied that day. Every time I hear the phrase 'He was like a kid let loose in a sweet shop', it reminds me of those days. At such shops, things hadn't really changed in 100 years. They had all the sweets in jars and served them in wee paper bags for as little as a penny. My particular favourites were soor plooms, pineapple chunks, kola kubes and black jacks but there were literally hundreds to choose from.

Apart from sweets, the Gorbals people were also passionate about ginger – fizzy drinks of all types – which came in glass bottles. The favourite flavours were Irn Bru, Tizer, ginger beer, orangeade and American cream soda, and once again there was a great assortment to choose from. To supplement our pocket money, we regularly went on searches for ginger bottles. The shops had a system whereby if you bought, say, a bottle of Irn Bru or Tizer, you got a couple of pennies back on the returned bottle.

A gang of us went to the big Celtic and Rangers games and made a few shillings picking up the discarded bottles and taking them back to the shops. But numerous shopkeepers in the Gorbals became wise to this. They devised a system whereby they put their own personal stamp on the bottles. If the stamp wasn't on them, you wouldn't get a penny. There were always plenty of empty bottles around, though, because folk in the Gorbals had an unquenchable thirst not only for alcohol but also for soft drinks. The ginger-bottle money was used to buy sweets or comics. *The Beano* and *The Dandy* often gave away a free gift, like a little plastic deep-sea diver. You placed the toy diver inside a lemonade bottle filled with water, turned the rubber screw top and he dived up and down.

We also went midgie raking: going through the middens in the back courts, to find ginger bottles or 'luckies' – discarded chattels that we could either make money from or play with. We found all sorts of stuff in the middens and older boys often sold their discoveries on for a profit. There were rumours that in posh places like Kelvinside the lucky might find priceless pieces of jewellery or art. If anyone spent a few hours raking the middens, they inevitably ended up looking extremely dirty. Their mother would shout, 'Oh naw! Ye've no been midgie rakin' again, hiv ye? Ya manky wee sod ye, ye're aw clatty. Ah'd better get the carbolic soap oot and gie ye a good scrub doon in the sink.'

Some of the Gorbals boys became expert at the art of midgie raking. They had a procedure to keep themselves spick and span, so as not to let their parents know what they'd been up to. They took off their jumpers, shirts and even trousers and hung them up on a wall near the midden. Dressed in only their underpants, they sieved through the rubbish. And if they needed a wash before going back to the house, they had a scrub down in the sinks at the toilets at Gorbals Cross, which always had an abundance of complimentary carbolic soap.

After one long hard day raking dozens of middens, I had a quick scrub at Gorbals Cross then headed back to the house hoping my mother would be none the wiser. I was also eager to watch the latest episode of *The Alfred Hitchcock Hour*. Like a good number of children at that time, I loved the suspense and thrill of the Hitchcock programmes. But halfway through, my nose began to bleed. The furious flow of blood wouldn't stop. My mother gave me a hanky then a towel to try to stop the bleeding but it went on and on. As the blood continued to flow, I thought it was ironic. It was like one of the spine-chilling episodes of *Hitchcock* – where would it all end? My mother called in a neighbour, who said, 'There's only wan thing fur it: you'll jist hiv tae put a cold key doon the boy's back. That usually does the trick wi nosebleeds. If no, you'll hiv tae call the doctor.'

But even after my mother put the cold key down the back of my shirt, the bleeding continued to get worse. Three hours later, they called for an ambulance and I was whisked off to hospital. The next day, my mother was holding my hand at the hospital bedside, crying. The doctor had told her that I had symptoms which were consistent with the initial stages of leukaemia. After a week of examinations and tests, some of Scotland's top medical experts were brought in. It was concluded that I had contracted what was considered a particularly odd and old-fashioned disease called Purpura, a disorder of the blood vessels. My nose continued to bleed until the doctors decided to let it clot naturally. With the help of drugs, they said, the condition would rectify itself.

I spent six months in various hospitals, including Yorkhill and

Glasgow Infirmary, where I was treated as a medical oddity. Doctors and medical experts from all over Britain were taken to my bedside. I heard one physician say that I was a classic case of a Gorbals child, having this rare viral disease. One nurse told me that the singer Tommy Steele had had the very same condition as a child. Other experts told me that in the past Purpura had been called 'the king's disease' because some famous kings in history had contracted it. Lying in a hospital bed beside other sick children, I was convinced in the early days of having Purpura that I wasn't going to live long; I vowed that if I ever made it back onto the streets of the Gorbals, I would enjoy every minute of my life.

After the longest six months of my life, the doctors at Yorkhill said I was well enough to go back home. But I was told that I could not do anything overly physical like playing football or running. I vowed never to rake the middens again, because a specialist had told me: 'We think that you picked up this virus after coming into contact with the millions of germs and microbes that lurk in the Gorbals middens. That is why you are such a medical oddity.' The middens had nearly killed me and I was used as an example to other children in the area. Their parents warned them that they would end up like me, seriously ill and in hospital for months, if they carried on with their midgie raking.

After I came out of hospital, one of the most alarming things to see was grown men raking the middens. They were there day and night going from midden to midden until they had made enough to buy themselves a drink. For many of these men, midgie raking became a profession, a full-time job. They were so desperate for a drink, they had come to the conclusion that it was the easiest way for them to earn money. What a way to end up.

The middens weren't the only health hazard. Coping with smoke fumes in the Gorbals was a major problem for many people. It was bad enough with the chimneys belching out black soot but that coupled with people smoking heavily in public places did not help any breathing problems you might have had. Less than a year before I got Purpura, I, like many other children, developed a bad case of bronchitis, which developed into pneumonia, and I ended up in hospital for six weeks.

I found the Gorbals, especially when all the coal fires were roaring in the tenements in the winter, a hard place to breathe. Also, most adults seemed to be blowing smoke in my face. Smoking was so big that the local shop in Crown Street, Lennox's, made a fortune buying and selling Embassy coupons. They were seen as a real alternative currency. Some people had the psychology that the more they smoked, the more possessions they would have. If they saved enough of the coupons they could buy almost anything they fancied from a catalogue. The Embassy Coupon Catalogue offered hundreds of goods in exchange for the coupons. People peered through it and puffed their way through thousands of cigarettes to obtain what they wanted. Whole houses were kitted out that way.

My pal Albert told me a story that put me off smoking cigarettes for life. He said he had just visited his uncle, who had been a big smoker, in the Royal Infirmary and was shocked to find him gasping for air on an iron lung. He noticed his uncle was wearing a big, flashy, expensive-looking watch and asked him where he got it. Barely able to reply, he struggled for breath saying, 'I bought it wi ma Embassy coupons.'

Chapter 8

UNIVERSITY OF LIFE

When we were at school, older people said to us, 'Ah'll tell ye whit, ye should enjoy goin' tae school, cause after that it's workin' fur a livin' and tryin' tae pay the bills. Enjoy it while it lasts cause schooldays are the best days o' yir life.' During the '50s and '60s, finding a local school, no matter what denomination you were, was hardly a problem. There was an abundance of schools in the Gorbals. There was St Francis in Cumberland Street, Hayfield School in Moffat Street, Wolsley Street School, St John's in Portugal Street, Greenside School in Cleland Street, Abbotsford School, Adelphi Secondary, Camden Street School, the fee-paying Hutchesons' Grammar in Crown Street and the first place I attended – St Luke's School in Ballater Street.

It was a rectangular-shaped building with dark-grey bricks and playgrounds to the front and back. The front was where the girls played and the back playground was for the boys. The teachers were a strange assortment. There was a demented spinster who had no hesitation in whacking the children with a thick leather belt. There were educated younger women who saw the school as a stepping stone to a better school and area. And there were the older, more experienced teachers who had seen it all but were happy with their lot in the Gorbals. The school had a typically Gorbals hard-but-fair atmosphere about it.

The parish priests lived in their own two-storey house in the school grounds. The pressure got to some priests and they turned to a glass or two of something strong. One particular priest hit the bottle for a while and he sometimes wandered into the playground to kick a football around with us. I always thought that priests were under a lot of strain. They had given their lives to the faith and not having a woman perhaps contributed to their finding solace at the bottom of a bottle.

The priests often came into classes and asked, 'Which of you lads would like to become an altar boy?' All of the boys stuck their hands up shouting, 'Me, father, me!' They were eager for the social prestige and economic advantages that came along with the position. Parents were proud if their son was selected to be an altar boy. It signified that you were considered a good Catholic boy and a cut above the rest. Also, the money was quite good. They got a few shillings for attending ceremonies such as weddings and funerals. It was funny watching the altar boys going into the chapel with their heads held high, faces scrubbed clean and hair combed to perfection, because although they looked like little angels, we all knew that they were really mischievous Gorbals street urchins and up to all sorts of tricks. Being an altar boy was a glamorous part-time job that we all wanted, but the majority of them were from devout Irish Catholic families, with their mothers and fathers going to chapel every week. My family wasn't in the same worshipping league. When the priest came round seeking new altar boys, I would stick my hand up but I was always ignored. One of the rejected boys joked, 'Ah've got nae chance o' ever becomin' the Pope if Ah cannae even get a start as an altar boy.'

Many of my friends went to Protestant schools. I was always puzzled by the difference between Catholics and Protestants. While Glasgow Catholics looked up to the Pope, Jock Stein and Celtic, local Protestants worshipped Rangers, the Sash and the Orange Walks celebrating the defeat of the Catholics by King Billy in Ireland in 1690.

We often followed the Orange Walks. The Orangemen wore dark suits and large bowler hats, proudly displaying their sashes, twirling their batons and playing pipe and drum music. There was the mandatory rendition of 'The Sash':

> Here am I a loyal Orangeman,
> Just come from across the sea,
> For singing and dancing,
> I hope I'll please thee.
> I can sing and dance with any man,

As I did in days of yore,
And on the Twelfth I long to wear
The sash my father wore.

They proceeded up the street for what seemed like an eternity. It was considered a major insult to them if anyone crossed the road while they were marching. This act was called 'breaking the ranks' and it sometimes resulted in the perpetrator being beaten up by furious Orangemen. The police often turned a blind eye to this. The marchers sometimes had missiles or coal ashes thrown at them by locals offended by their anti-papal stance, but they stood up to the pressure with obstinacy and determination and carried on marching, believing they had a God-given right to do so.

Alex did not go to either a Catholic or Protestant school in the area, even though he lived in Thistle Street only a block away. Every morning, he stood outside of his close and a little grey van picked him up and transported him to a special school on the outskirts of the city. Street-wise, he was a genius, but to Glasgow's education officials, he was considered backward. I thought Alex was the flyest guy in the world, having what looked like a taxi taking him to school. But my mother, like hundreds of other mothers in the Gorbals, often warned, 'You'd better start behavin' yirsel at school, otherwise you'll end up in the grey van. Dae ye want aw yir pals laughin' at you when the grey van picks ye up every morning? Everybody will be thinking ye're no right in the heid.' The threat of ending up being taken to a special school in a grey van with other children who were considered mentally backward was usually enough to keep us on the straight and narrow.

The teachers at St Luke's had their own inimitable teaching techniques. For example, if we were learning the twelve times table, the teacher would pick someone out of the class and shout, 'What is seven twelves?' If the pupil didn't get it right, he or she would receive 'one of the belt'. This meant getting struck across the palm with a leather belt. The resultant pain meant that we tended to learn the times tables rather quickly.

The headmistress was a small, thick-set woman who was an authoritarian character. She was very much a product of the old

75

education system. My belief was that she had it in for me. Any time she was around, I tried to keep a low profile. She had told other teachers that I had a big chip on my shoulder and needed a lot of sorting out. I was aged around ten when she saw me combing my hair in front of a mirror near her office. She was furious; her face was as red as a beetroot. She shouted, 'Boy! What do you think you're doing? How dare you do that here in full view of everyone! Did no one bother to tell you that it is the height of bad manners to comb your hair in public? You have given me no alternative but to belt you.' I was hauled into her office and given three strokes of the belt, which was extremely painful. I thought, as I was receiving the punishment, that there were robbers, con men, fly men, drunks, pickpockets, gangsters and even murderers out there in the streets of the Gorbals up to all sorts of no good, and here was I getting belted for combing my hair. If I didn't have a chip on my shoulder before, I certainly did after that.

At playtime, gangs of boys kicked a ball about, screaming and shouting with the odd square go taking place. Chris got into an argument with another boy and asked me for a loan of my snake belt. At that time, most boys in the Gorbals and the rest of Glasgow wore a snake belt, a small elastic belt with an S-shaped metal buckle. It had proved to be very versatile in fights. I unfastened the snake belt and handed it to him. He promptly got into a fight with the boy and hit him over the head with it. The boy put his hand on his bloodied head, running about crying and shouting, 'Help me! Help me! He's tryin' tae kill me!' We were soon before the headmistress receiving the maximum punishment she could hand out – six of the belt each. Chris had been the main culprit but I was considered equally to blame because, after all, it was my belt that had caused the damage. The snake-belt incident meant huge black marks against our school records and it would be a deciding factor when it came to which secondary school we would eventually go to. St Bonaventure's was for the rough-and-ready types and dunces, Holyrood Secondary for the intelligent boys who behaved themselves.

No matter how many times I got the belt, I thought St Luke's was probably the best school in Glasgow. There was a street song called 'Oor Wee School'. It summed up the atmosphere and the love the pupils had for their own particular schools, in the Gorbals and in the rest of Glasgow:

Oor wee school's the best wee school,
The best wee school in Glesga.
The only thing that's wrang wi it,
Is the baldie-heided master.

He goes to the pub on Saturdays,
He goes tae the church on Sundays,
He prays tae the Lord tae gie him strength
Tae batter the weans on a Monday.

A few blocks away in Crown Street, and after 1960 in the nearby Crosshill area, there were pupils who belonged to a completely different world from oor wee school. Their parents paid good money for their children to be educated, and the pupils of Hutchesons' Grammar School attended in their smart blazers, white, ironed shirts, ties and leather briefcases. Local boys often shouted insults at them and the girls chanted at the schoolboys:

Tell-tale tit,
Yir mammy cannae knit,
Yir daddy cannae go tae bed
Withoot a dummy tit!

These kids were the middle-class elite of Scotland, the sons of wealthy doctors, lawyers, dentists, chemists, accountants and businessmen, but Gorbals kids never felt second rate. They might not have privileged tutoring, but what they learned on the streets was priceless. Some of the older local characters often told us, 'Whit ye learn aboot life here is better than payin' tae go tae a private school wi a load o' hawf-boiled toffs. It's something money cannae buy. You've got the best education of aw – the University of Life.' Alex replied, 'Aye, the only problem wi that, mister, is most o' the people who went tae the University of Life failed there as well.'

Chapter 9

PLAYTIME

Although they were dirty and the ground was usually covered in puddles, the back courts of the Gorbals tenements made terrific playgrounds. They were ideal for acquiring a variety of physical abilities. Leaping from wall to wall became not only a skill but a street art form. Some of the walls had sharp iron railings and we all knew it was dangerous but it did not dampen our enthusiasm. Miraculously, none of us ended up being impaled but there were rumours that the nearby Caledonian Road Graveyard was full of children who had been skewered while playing. The worst that happened to most of us as a result of leaping the back-court dykes was a skint knee, which was treated with zinc ointment and a bandage. Our mothers would watch our antics from the comfort of their window ledges, often shouting things like, 'Dae ye want tae end up in the Royal Infirmary?' We usually just carried on as if we hadn't heard them (which we called 'slinging a deafie').

The art of 'dreeping' from the walls was very important to many back-court games. The best way to get off a wall was not to jump to the ground but to dreep. An older boy told me when I was wee, 'Aw ye've got tae dae is lie doon on yir belly and haud on tae the wa' wi the tips o' yir fingers before ye let go.' I took his advice and found that by dreeping I always landed safely and softly on my feet. Sometimes there was even the luxury of an old mattress on the ground to cushion your landing.

During the long summer days, playing a game of jorries, or marbles, was a favourite pastime. Children played for hours on the numerous stanks – drain covers with holes. These games could go on all day and the victor was the one who ended up with the most jorries. Some boys had hundreds of marbles in their collection and everyone had a special marble, a lucky jorrie, which was believed to win a major game against any opponent.

Other street games included cowboys and Indians, which always led to tying someone up and holding them prisoner until the cowboys or Indians managed to release them. Another involved making holes in two tin cans, putting string through them and using them as stilts. Meanwhile, the girls walked about with their mothers' old high heels on, feeling very grown up.

The lassies had their own games. There were always crowds of them in the back courts and the streets playing at stoat the baw. They bounced a rubber ball between their legs against a wall and then, with great dexterity, caught it. Skipping was also popular. A large length of rope was held at either end and swung around in a large arc (this was called 'cawing') while the girls skipped up and down.

Another game played in the back courts was called 'kick the can'. It involved placing a tin can in a circle before the players hid in the back-court or nearby closes. It was like hide and seek. One person was left standing by the can and it was his or her job to find the hidden few. The danger was that if the seeker wandered too far from the can, a hider would run up and kick it. This meant that anyone who had been caught was released and the game had to start all over again. If someone was caught, the seeker chanted out: 'Come oot, come oot, wherever you are, ye're no het!' or:

> Come oot, come oot,
> Wherever you are,
> The game's a bogie,
> The man's in the lobby,
> Eatin choc'lat buscuits!

Kick the can was actually very good training for hiding from people when you were being chased. If you played the game often enough, you knew every nook and cranny in the Gorbals. A more mischievous game was 'kick the door, run fast', which involved going to the top of a tenement, kicking the door and then running down each flight kicking every door on the way. But residents got wise to this. By the time we ran out of the close we usually had a pail of water, or even a chamber pot full of pish, thrown over us by the irate victims.

Dummy fighting was one of the most violent games you dared to play. It could be dangerous, because often the dummy fight turned into a real fight and there would be blood everywhere. One boy pretended to attack you with a belt and you would pretend to fend him off. But one thing usually led to another and then it would escalate. Kids ended up with stitches and even scarred for life after having dummy fights. My father warned me, 'There's nae such thing as dummy fighting. Ye end up getting a doing. Real fightin' is fur the real people and dummy fights are fur the real dummies.'

Although the police had a fair workload in the Gorbals, some of them took to booking groups of children for playing innocuous little games. They were not content with arresting criminals but wanted to book boys and girls for playing football and other ball games in the street. This was the worst public-relations exercise they could have carried out. It caused a lot of resentment and the children grew up detesting the police, a feeling that stayed with them all of their lives. Twenty of us were booked one day for playing a game near Lawmoor Street. Somebody's mother, who was leaning out of the window, shouted to the two officious policemen, 'Why don't ye leave the weans alane and catch some real criminals like bank robbers or murderers? Ye should be ashamed of yirsels picking oan weans like that.' Seeing the big red-faced woman in action made the two policemen look embarrassed and they both made off rather sheepishly. We cheered the woman on. She was right. Surely the police had better things to do than pick on children for playing street games? We were hardly big-time criminals, although, ironically, some of us would be in the future.

Along with the streets and the back courts, swing parks were popular places to congregate. Competitions were held to see who could go the highest on the swings. This involved great coordination and strenuous arm and leg movements which were really quite hard to master. Some of the children performed what looked like death-defying feats; they got to tremendous heights on the swing and then leaped off, landing on their feet without a bruise. To keep order in the swing parks there was a warden, the parkie, who made sure that the games did not get out of control. If anyone got out of order, the parkie would chase them out, shouting, 'Get oot o' here and don't come back until ye learn how tae behave yirsel.' This usually led to far more mischievous adventures in the streets, like leaving a can of urine against somebody's door, knocking on the door and then running off, ensuring that the offending pish went all over the person's lino or carpet. A crueller trick was to put excrement in a paper outside of a door and then set it alight. We then knocked on the door and ran off. Seeing the flames, the alarmed resident would stamp on the offending article, ultimately covering their shoes in the muck.

When Guy Fawkes Night came around, the preparations for the big event became one long game. Before 5 November, weeks were spent collecting firewood from all over the Gorbals in order to erect the biggest bonfire imaginable. Because there were so many old buildings being demolished, there was no shortage of firewood. It became a competition: if someone built the biggest bonfire, they became a big shot amongst their pals. 'A Penny for the Guy' was the easiest way to earn money and the best places for that were outside of the pubs. The boys picked on men who had a really good drink in them, as they were the most generous. The men threw handfuls of change, saying that it reminded them of when they were youngsters and 'didnae hiv a damn care in the world'. Some of the boys got out of control and threw penny bangers into pubs, giving the drinkers a fright as the fireworks exploded while they were having a quiet pint. A boy I knew in Crown Street was rushed to hospital after one of his pals stuck a banger in his back pocket for a joke; he spent months in hospital being treated for skin burns.

Hallowe'en was another time for the children of the Gorbals to go a bit daft. Large turnips were carved into ghoulish faces and lit with a solitary candle. The boys and girls went round doors and stood outside of the pubs with blackened faces guising – singing a song for money or an orange or apple. It was simple: if people were too poor to give you money, they at least ensured they had a piece of fruit to give you for your trouble.

While playing, the girls and boys sang the most wonderful street songs. The sweetness of the melodies resounded through the air, especially on warm summer nights. The Gorbals children had a great propensity for memorising such songs, which had been passed down from generation to generation. Most of these songs were vulgar but extremely funny. One favourite was 'Oh, Ye Cannae Shove yir Granny aff the Bus':

> Oh, ye cannae shove yir granny aff the bus,
> Oh, ye cannae shove yir granny aff the bus,
> Oh, ye cannae shove yir granny,
> Cause she's yir mammy's mammy,
> Oh, ye cannae shove yir granny aff the bus.
>
> Singing, I will if you will so will I,
> Singing, I will if you will so will I,
> Singing, I will if you will,
> I will if you will,
> I will if you will so will I.

The children on the street were never short of humorous songs to sing and they lightened up what could be an austere existence. Even years later, these words remain carved into the consciousnesses of people who lived in the Gorbals at the time. Many of the rhymes were about mothers:

> If you should see a big fat wumman
> Standin' in the corner hummin,
> That's ma mammy!

Another 'mammy' favourite was:

> My maw's a millionaire – would you believe it?
> Blue eyes and curly hair,
> Sitting among the Eskimos,
> Playing a game of dominoes,
> My maw's a millionaire!

One of the earliest street rhymes I can recall learning was:

No last night but the night before,
Three wee monkeys came to ma door,
One wi a trumpet and one wi a drum
And one wi a pancake stuck to its bum.

Singing was not confined only to the youngsters. Men who had fallen on hard times often toured around the back courts belting out a hotchpotch of ballads. They would stand in the middle of the back court and serenade all who cared to listen. People hanging out of their windows then put a few coppers into a folded piece of paper and threw it to the singer. I saw one such man entertaining the back-court audience for more than an hour and he made a grand total of tuppence. He looked at the meagre sum he had earned, shook his head and then said, 'There's nae bloody money in the Gorbals. Things are so bad, it's me that should be throwin' them money.' Some of the older children had a cruel streak in them. They would get a pair of pliers and put a penny over a gas fire until it was red hot before throwing it out of the window. The singer would pick it up and emit a shriek of pain as the coin stuck to the flesh on his hand. Another back-court singer put on a brave face after being burned by a hot penny. When he picked up the coin he let out a yell of 'Ya wee f****n' bastards, yis!' before quickly putting his sizzling hand in a puddle to cool it down and saying, 'Ah suppose ye've got tae expect this, it's all part o' the job because there's nae business like show business!' For these men, there was no great fortune or fame but the show had to go on.

The alternative to hanging about the streets of the Gorbals was to start behaving and go to one of the many youth clubs that had sprung up. They were seen as a genuine attempt to stop youngsters getting into trouble, but one of the most successful clubs was run by a man who, rumour had it, was a 'touch-up merchant'. I made several visits there and I noticed he was playing with the boys and touching them all the time and even resorting to tickling them. I thought that if he got his fun by tickling young boys there was bound to be something wrong with him. The local Protestant boys joined the Boys' Brigade, which was supposed to promote good habits, including obedience, reverence, discipline and self-restraint. A man from Glasgow, William Alexander Smith, had started it up in 1883. It had a military tone to it, with drills, dummy rifle exercises, various sports and a pipe band. The Protestant kids loved it and it became a major part of their growing up. There were many other options including the Cubs, the Scouts, church clubs, football clubs and pipe bands.

However, we found them all a bit regimented for our liking; it was bad enough being at school, having to put up with the discipline of rules and regulations, without having to adhere to similar ones in youth clubs. We all got a bit exasperated when one guy at another club tried to force us to learn how to play the bagpipes. We thought that no Gorbals tough guy would be seen dead with a set of bagpipes; that was Highland bampot stuff. When I told my father, he couldn't stop laughing. He said, 'Dae ye know the definition of a Scottish gentleman? Somebody that can play the bagpipes . . . but disnae! He must be a brave guy trying to teach you wild young guys the bagpipes. Remind him it's the Gorbals he's in, no Oban.'

The Gorbals attracted all sorts of creative people seeking inspiration. There were poets, comedians, musicians, playwrights, songwriters, novelists, photographers, artists, philosophers and academics. They all made visits to the Gorbals because it was a place that could conjure up a million different images and ideas. One of the best ways to observe all of human life was simply to stand at the mouth of a close. A gang of us were passing by a close in Crown Street when we spotted a little man with a goatee beard wearing a raincoat. He was clearly soaking up the Gorbals atmosphere. One of the boys recognised him as folk singer, songwriter and comedian Matt McGinn. He had made a name for himself on the folk scene writing songs and telling outlandish stories about Glasgow life. McGinn considered the Gorbals to be an ideal place for inspiration. We started a cheeky conversation with him but he had fast patter and held his own. Like us, McGinn had had a tough childhood, before working in the shipyards and later becoming a primary school teacher. He had a great affinity with the local children and in 1964 he was behind an idea to set up an adventure playground in the heart of the Gorbals. The site was a spare piece of ground near some tenement slums. Glasgow Corporation had been using it as a dumping ground for broken concrete sewage pipes and leftover sand and gravel. The Corporation planned to smarten the place up and make a small park there, but McGinn managed to persuade them to leave the place pretty much as it was, simply clearing away the junk and putting up a fence around the area. He also encouraged them to erect an old Nissen hut on the site so the children could play there on rainy days. Outside, they ran wild on rope ladders and old car tyres used as makeshift swings. It would never have got past today's safety regulations but it provided a haven for hundreds of Gorbals children. Inside the hut, the walls were plastered with newsprint rolls, donated by the Glasgow *Evening Times*. These were used for graffiti training; the Gorbals children wrote and drew the most obscene messages imaginable before the newsprint was taken down and replaced with fresh paper. Most of the boys I knew thought of themselves as a pretty hard bunch but when they visited the new playground they quickly discovered that the boys and girls who played there were much tougher and scruffier.

The adventure playground, known as 'the venny', was surrounded by poles and wire mesh. It reminded us of one of those films about prisoner-of-war camps. The place had its regulars and they took exception to strangers coming into their bit of territory. Everyone I talked to at school or on the street had been beaten up while playing there. On one occasion, I had only been there a few minutes when four shabby-looking boys approached me and one of them said, 'Who you lookin' at?' Before I had time to reply, I was kicked and whacked about the head, but luckily enough I managed to run off.

McGinn persuaded fellow folk musicians to give a monthly concert at the adventure playground. The 50 or so scruffy children who turned up for the concert in turn taught the visitors the dirtiest songs they had ever heard. The singers were often threatened. The boys shouted things like: 'My big brother is gonnae kick your heid in when he gets oot o' jail.' Some of them even tried to set up a protection racket, saying to a musician, 'Gie's two bob to look after yir guitar, mister.' When the musician refused the offer by replying, 'Thanks, pal, but I'll look after it myself. It won't be leaving my side.' He was greeted with the quick retort, 'Awright, mister, but a hawf brick wouldnae smash intae it if Ah wis looking after it.' The *Evening Citizen* reported that a young American volunteer, Jane Finlay, had difficulty understanding the local language: 'One child asked me to "geez a coaxie [give me a piggyback ride]". It took me a while to figure out what he was saying.'

When the venny was opened, it was described as a play paradise. It was staffed by a full-time youth leader, two part-time play organisers and voluntary helpers. Play organiser McGinn told the *Citizen*, 'My job here is not to tell the children to do this or that, but to be there if they want help. Another important, if unofficial, job is to give them lights for their fags.' In 1966, Glasgow councillors inspected the playground and promptly cut its £1,000-a-year grant. They were not impressed by the wild and unconventional set-up. The playground was threatened with closure and the locals protested. They claimed it had cut down on vandalism because the kids now had an outlet for their pent-up energy and aggression. Some 500 kids sporting homemade banners marched to George Square to present a petition to the Lord Provost. As a result, the Corporation provided more funds to keep the venny open.

There was no doubt about it, though, the venny catered for very wild youths. A joke that was going about at the time about the kinds of kids who were venny regulars went: a salesman knocks at a tenement door and it is answered by a 12-year-old boy with a cigar in one hand and a bottle of Lanliq in the other. The salesman asks, 'Excuse me, son, is your maw and da in?' To which the boy replies, 'Dis it look like it?'

The poorest children at the venny could be easily identified because they wore wellies most of the time. Wellington boots became a byword for poverty. Many parents could not afford to buy their children shoes, so they were destined to live out their childhood in wellies. They were good for playing in puddles and tramping through the dark, dirty back courts but it was a ridiculous sight seeing a child wearing wellies at the height of summer. It was as though they had been welded into them.

Two women were discussing one such child when one disclosed: 'Wee Jenny wis in terrible pain wi her legs an' she wis hardly able tae walk.'

'So Ah heard,' said the other, 'but Ah hear she's still no right efter the operation.'

'Oh, whit kinda operation did she huv?'

'They cut three inches aff the tops o' her wellies.'

Because we had so much interaction with each other, children like myself growing up in the Gorbals in the '60s never felt lonely. We always had the company of other children. I occupied myself by playing games all the time and strolled to school and back with my pals, with an abundance of mischievous repartee and patter on the way. These days, circumstances have changed. It's disheartening to see so many children being chauffeured by their parents to and from school. Lots of them, when they get home, head for their rooms to play computer games. Back then in the Gorbals, even in the house I was never on my own because I shared my bedroom not only with my brother but with my mother and father. Indeed, at one point, we all shared bunk beds. Today, countless children will never experience the camaraderie that once survived in neighbourhoods like the old Gorbals. And, as my mother used to say, 'Ye cannae beat a wee bit o' company, because loneliness is the saddest state in the world.'

Chapter 10

SCRAMBLING AND GAMBLING

I f we were ever in need of money quickly, going scrambling was one
way to make a few bob. In those days, when a couple got engaged they
would save up their small change to toss to local children in the street
on the day of their wedding. Sometimes the father of the bride threw
the coins in the air as the girl left her house for the church, sometimes
the groom started the scramble outside the church after the wedding.
Often, on a Saturday morning, we'd go round the chapels finding out
when and where the weddings were taking place so that we could fit
in as many scrambles as we could. What you got was mainly pennies,
threepenny bits and tanners but more generous grooms threw in an old
silver tanner, the odd two-bob bit or half a crown, so some weddings
proved to be more lucrative than others.

As the newlyweds came out of the church, the children would be
like a pack of wolves waiting for their prey. The groom would shout,
'Scramble!' and throw the cash into the air. The scuffle that followed
could be very competitive and it helped if you were fast, big or preferably
both if you wanted to end up with few scrapes and bruises and plenty
of coins. The police and health officials wanted the tradition of the
scramble banned. They warned that it could end in a tragedy, especially
as more and more cars were being used at weddings. However, most
adults loved to see the sight of all the young people scrambling for the
money.

One time, Alex and I joined a scramble near Gorbals Cross and the
other children were so young and puny, it was easy money. At the end of
all the pushing, shoving, shouting, bawling and screaming, between us
we had around two pounds and a couple of minor bruises. To celebrate,

we thought we'd go for a fish supper in nearby Eglinton Street. But by the time we got to Florence Street, we realised we were being followed by a gang of much older and bigger boys. They grabbed hold of us and demanded, 'Gie's yir scramble money otherwise we'll kick yir heads in. We've been watchin' ye since Gorbals Cross so hand over the cash noo.' But just as I was about to give them my scramble earnings, I saw Eddie, a fly man I used to talk to who stood every day outside of the bookmakers in Crown Street. I shouted, 'Hey, Da! Help me!' It was an old trick but it worked every time.

The fellow made towards us and the boys ran off. Relieved, we began to laugh, and as Eddie approached us we realised that he was half cut. We told him what had happened and walked back towards the safety of Crown Street. He declared, 'Hey, boys, Ah've just hid a great tip on a dead cert. It's 25–1 and it's got nae chance o' gettin' beat. Ah had a good bevy at the Mally Arms wi an old pal o' mine who's got inside information at Ayr Racecourse. But the problem is Ah've done my lot in the boozer. Gie me yir scramble money and Ah'll make us a fortune.' He sounded earnest enough, but maybe the whisky had clouded his judgement. On the other hand, he never went to work and always had the money for a drink, so maybe he knew what he was talking about. We said we wouldn't give him all our money but we would risk a pound on the 25–1 shot.

'Don't worry, we're ontae a winner, ye cannae go wrong wi this wee smasher of a horse, it'll gallop hame,' he said, clutching a handful of our coins. He went into the betting shop. We waited patiently, talking excitedly about what we were going to do with our share of the winnings. He came back out a few minutes later with a face as long as that of the horse he had backed. He said, 'Ah cannae believe it! The wee bastard was leading all the way and it jist got pipped at the post! But listen, if ye gie me that remaining pound o' yir scramble money, Ah've got another horse that's running at 50–1. My inside information tells me that it's a better bet than the last wan. It'll romp it, nae bother at aw.'

I was thinking a number of things: did Eddie the fly man take us to be simple-minded young mugs? Did he really think we came up the

Clyde on a banana boat? Or was he in fact telling the truth? He said, 'Hurry up, boys! Ye've got tae make yir minds up. Ye'd be right bampots no tae put yir money on. The race is startin' in a couple o' minutes.' We handed over our remaining pound, still more or less convinced that this half-drunken fellow could do us no wrong.

By this time, the ferocious Glasgow rain had begun to pour down and we were getting soaked to the skin waiting for the result. It was a good ten minutes before fly-man Eddie came out again and when he did, he had an apprehensive look on his face. He said, 'Ah'm sorry tae let you doon, boys, but the horse let me doon even worse. It turned out to be a right donkey. It's jist no ma day. Ah should have stayed aff the bevy, cause it's nae use bettin' after ye've hid a drink. It's whit aw the losers dae and Ah should have known that before gieing ye the tip.'

We shrugged our shoulders and walked off vowing revenge. Alex said he'd get his big brother, a leading member of the Young Cumbie gang, to 'kick f*** oot o' Eddie'. But we were halfway up Crown Street when Eddie shouted to us again. We turned back and he burst out laughing like a lunatic. He was waving fifty-odd pound notes in his hand. 'You stupid wee eejits,' he said, 'Ah wis only windin' ye up. Dae ye think I would mess ye aboot after giving ye ma word? The horse wis magic, couldnae go wrong. She flew it. Who says bevy and betting disnae mix?'

He gave us half of the winnings, more than £25. We must have been the wealthiest children in the Gorbals that afternoon. Poor little rich kids – how we dearly loved our lives on the streets!

The punters at the bookies were always full of patter. One wee man, Barney, had a long-running joke with us. Every time we asked him how he had got on at the bookies he said, 'Ah backed a horse at ten tae wan and it never came in till quarter to three.' No matter how many times he said it, it always made us laugh. Every time we met Barney he also told us his latest betting jokes. Our favourite was: 'A big Irish fella walks intae the bookies in Crown Street and puts a pound oan an accumulator. Every wan of his horses comes in and he is due £10,000 fae the bookie. But the bookie says to him, "Ah'm sorry, Paddy, Ah've hid a bad day, can ye wait until the banks open on Monday and Ah'll gie ye the £10,000 then." And Paddy replies, "Look, if ye're gonna mess me about, give me ma pound back.'"

There were all sorts of gamblers going about the Gorbals. I saw men bet on the stupidest of things, like the speed of competing raindrops running down a windowpane. Some days, especially in the summer, dozens of men gathered in Florence Street for a game of pitch and toss. A man stood in the centre of the crowd and threw two pennies into the air while the crowd bet on the outcome: either two heads or two tails, or odd or evens. This game had been going on in the Gorbals streets since anyone could remember. There was always someone on the lookout for the police. It was a strange sight, like taking a trip back in time, as many of the older men had been playing pitch and toss since they were boys in the 1920s. A man could lose his week's wages or win a small fortune according to how the coins dropped. Some families lived well after a game, others starved.

A lot of folks were also dog fanciers and made and lost fortunes at the dog track at Shawfield in Rutherglen. There was usually a good tip going round about a dog, especially if it came from the area. Numerous stray greyhounds wandered through the Gorbals streets and back yards after being discarded by their owners for not being fast enough in chasing the rabbit.

Even the bingo fanatics had their betting stalwarts. One seventy-five-year-old woman, known as Auld Aggie, marked six bingo books at a time, a feat that younger bingo-goers could never have achieved. We asked her one day what her secret was. She replied, 'Ah've brought up twelve weans in the Gorbals and Ah hiv hid tae keep an eye oan every single wan o' them. So keepin' ma eye on six bingo books at a time is a doddle compared tae that.'

Chapter 11

RELICS OF A BYGONE AGE

When I was growing up, a few local people had cars but the majority travelled on the numerous Glasgow Corporation tramcars and buses to get around the city. The trams were a culture unto themselves, with the female clippies keeping the passengers under control and in order by shouting, 'Come oan, get aff!' or, as I heard a clippie shout one day, 'If ye're no gettin oan, get aff, and if ye're no getting aff, get oan!'

I heard a great line one day when a passenger asked for a single to the city centre. The clippie replied, 'This caur isnae gaun tae the city centre.'

'But it says the city centre oan the front.'

'Mister, it says India oan the tyres and we're no gaun there either!'

Another classic was the response to, 'Ur there any seats upstairs?'

'Aye, but there's arses oan them.'

There was a street song about working as a clippie on the trams:

> Fares, please, fares, please,
> You'll always hear me say
> As I go up and doon the tramcaur every day.
> Oh, Ah work fur the Corporation,
> You'll know me by my dress,
> Ah'm Lizzie Macdougall fae Auchenshuggle,
> The caur con-dut-er-ess.

But sadly that culture was to end on 1 September 1962. A quarter of a million Glaswegians gathered in the streets in torrential rain to wave goodbye to the last of the trams. The much-loved Corporation trams disappeared from the streets forever, much to the dismay of the general public and the music-hall comedians. One of my neighbours in Crown Street, Mr McLean, a retired tram driver, missed them so much he began taking his holidays in Blackpool, where there was an abundance of trams. He said, 'Before Ah retired, Ah drove the tram every day for mair than 20 years. Ah miss them. They say the fancy new buses are the future but they hivnae got the same atmosphere aboot them. That's why Ah'm aff tae Blackpool – it's the only place Ah kin get a caur ride.'

Another aspect of our lives which was soon to be a thing of the past was the ragman. Every now and again the ragman would come with his horse and cart. He patrolled the streets, blowing a trumpet and shouting, 'Any auld rags? Any auld rags?' People hurled bundles of rags out of their windows and children ran to their flats in the tenements to fetch old garments so that they would get their reward of a colourful balloon. A balloon was a good deal to a child for a pile of rags. I saw a wee boy in Florence Street get furious when his mother refused to give him anything to exchange. When her back was turned he took off his new woolly jumper and gave it to the ragman for a big yellow balloon. When his mother found out, she went berserk and chased the ragman up the road shouting, 'Ya robbin' swine, ye! Gie me back ma boy's jumper! It cost me twelve and a tanner fae the knitwear shop, it's worth far mair than wan o' yir daft balloons!' A standing joke was if you saw a pal dressed in shabby gear, you said, 'You'd better pull yirsel together, the ragman's coming.'

Some impoverished people would take a bundle of their old clothes to the rag store in Ballater Street and try to exchange them for money. The shop was situated under a railway bridge and looked dark and mysterious from the outside. Inside, though, it was like an Aladdin's cave, with all sorts of goods. People could pick up a half-decent second-hand cooker, bed, wardrobe or chest of drawers cheaply and the owner usually had what they wanted. Many newcomers and immigrants furnished their whole houses from the rag store. My father went in there once after a few bevies to have a browse and ended up buying an old decrepit-looking rocking horse for a few shillings. Surprisingly, we kept it for quite a few years. The proprietor didn't charge a fortune like the big showrooms in the city centre did. He surmised that most of his stuff was junk and therefore charged junk prices. The boys never tried stealing anything from there because the guy had a terrible temper. The rumour was he had once bashed someone's head in with a poker after he caught them stealing. He gave us a bit of friendly advice: 'You boys are welcome tae hiv a look aroon' here but if Ah catch ye stealin' anything, ye'd better like hospital food.' Not being big fans of hospital food, we quickly got the message and confined ourselves to just looking around at all the items on display. Besides, we reckoned that you would have to be the lowest of the low to stoop to shoplifting from the rag store. The prices were so low the man was almost giving the stuff away.

Another throwback to the olden days that survived was gas lighting. Even by the late 1960s, many of the tenement closes in the Gorbals were still lit with gas. Some flats also had gas lighting, which was like taking a trip back to Victorian times. I remember visiting such a flat in Thistle Street and just thinking it was incredible that people still had to put up with such living conditions. The Corporation employed lamplighters who went into the closes every night with a lit gas torch to illuminate the lamps. They took great pride in their job, carrying little ladders over their shoulders as they trudged from close to close.

Sometimes they left their ladders unattended and kids would hide them in places like a nearby midden, much to the fury of the lamplighters. The ladders were an essential part of their job and without them they could not operate. One night my pals and I hid an elderly lamplighter's ladder and he became furious with us and very emotional. He was at his wit's end and shouted at us, 'Whit hiv ye done wi ma ladder? Ah've hid it since Ah got the job after leavin' school. It's o' great sentimental value to me, like one o' ma family. Gonnae please gie's it back!' We felt so sorry for the man that we quickly directed him to the midden where his ladder was found safe and sound. In 1971, in North Portland Street, 12 of Glasgow's oldest lamplighters witnessed Lord Provost Sir Donald Liddle light the city's last gas street lamp for the final time. It was the sole survivor out of the original 24,000. Many of the antiquated gas lamps had been stolen by fly men and punted to dealers who shipped them all over the world.

Like the gas-lighting fellows, coalmen were always up and down the tenement stairs. Most of them were hard drinkers but day in, day out they were able to put large heavy bags of coal on their backs and deliver them to flats in the tenements. It wasn't a job for the weak or the workshy; the coalmen often had to climb three or four flights of stairs with the huge bags. They arrived in the street on a horse and cart shouting 'Coal!' to let people know they were coming. Then they climbed the tenement steps, still shouting, so that the residents knew to open their front doors to let the coalman through to their bunkers.

The posher folk, in their wally closes with patterned tiles and stained-glass windows, tended to be treated a bit differently from the normal working-class customer. Their closes were always freshly scrubbed and everything was immaculate. They did not want the coalmen to make a mess. The coalman treated them with the respect they expected and did not shout for them to open their doors. Instead, he waited until he got to the door and pulled the polished bell knob. Then he'd wait with the heavy, wet bag pressing against the nape of his neck until the door was eventually opened. The customer would then carefully lay out a newspaper in the lobby as a precaution against the coalman's tackety boots making a mess on the spotless linoleum. Also, the toffs never had the coal delivered on a rainy day in case the dust made a mess when it was dumped into the bunker.

Jimmy the coalman had been delivering in the Gorbals for years and he looked as though he was getting on a bit. He had thick grey hair and a face full of lines but he could pick up a 112-lb bag of coal and fly up a three-storey tenement with the greatest of ease. It was as if the bag were a minor irritation that he had got used to over the years. His hands weren't particularly big but they were strong and firm. Once, when he shook my hand, it was so painful I thought that he had broken my fingers. His hands were often covered in cuts, caused by scraping coal from his cart and catching lumps mid-air as they fell from his bag. Jimmy earned quite a bit of money but spent most if it, he said, on his three favourite vices: malt whisky, fast women and slow horses. If he had time, especially on a rainy day, he'd stop in a close, light a Woodbine cigarette and tell anecdotes that were bound to capture your attention. 'Ah deliver coal to all sorts o' hooses,' he said to us boys. 'No only here but to the posh hooses in Kelvinside. In places like that, they never ask ye if ye want a cup o' tea because they don't want ye tae dirty their hoose. But the Gorbals is different. If Ah took a cup o' tea fae everybody that asked me, I'd never get any work done.

'Ah've got ma regular customers and aw o' them are good payers. But ye dae get the odd wan that's hit hard times and that can be difficult. I sometimes get a woman asking if she can pay on her back but Ah wid never agree tae that. I want money for ma coal, no a ride. Besides, if the missus found out there wid be blue murder.'

It did surprise me that some women offered sex in exchange for a bag of coal and that some of the other coalmen took advantage of the offer. Apparently, they saw it as a bonus for working so hard. The gossips would say things like, 'That wee Doris, her man's in jail, her weans are in rags, she's got nae food in the hoose and no a penny tae her name. So how come she's got a big roaring fire aw the time?' No one answered the question directly but there were a few grunts and knowing nods.

The coal fire was essential to combat the severe cold in the tenements. It was trial and error getting a good fire going from scratch. First of all, you had to rake out the ashes and put some rolled-up newspaper in the grate, then you added sticks, placed the coal lumps on top, lit a match and saw how it went. But the wet, windy nights made it difficult and you had to 'draw' the fire with the help of a shovel and a broadsheet newspaper like the *Evening Citizen*. I discovered an even quicker way to get the fire roaring. The corner shop began to sell firelighters for a few pennies. They were white oblong bricks that smelt of petrol and got the fire flaming in no time. But the more traditional fire-makers looked down on this habit, as it was considered cheating just a bit.

Once the fire was roaring, we all sat round it and it became a focal point for the family's philosophising. In front of the glowing flames, we discussed the problems of life, brought back memories and contemplated our hopes for the future. I remember sitting one night with my granny in front of a roaring coal fire when she stirred it with a poker and said, 'If ye look closely and long enough, ye can see yir whole life in a coal fire.' I knew what she meant. The coal fire was like a work of art, with thousands of images in the flaming embers.

Younger women avoided going too near the fire. If they put their legs close to it for too long, they ended up with 'tinker's tartan', which were the marks made by the heat of the fire; they did resemble bright-red tartan. It was certainly no asset but some of the older women displayed it proudly as they walked through the streets. It seemed as if, to them, tinker's tartan was a symbol of womanly maturity.

Apart from the coalmen, another source of fuel was the briquette sellers who patrolled the streets with their barrows, which they hired for a couple of shillings a day from a firm in a lane just off Gorbals Street. They punted not only coal briquettes but such things as fruit and veg and fish. Many of them made a nice living out of pushing their barrows up and down the streets. One local fellow started off with a single barrow selling fruit and veg, then progressed to a string of them before acquiring shops all over Glasgow, making himself a wealthy man in the process.

The trams, the clippies, the ragmen, the lamplighters, coalmen and barrow boys, all of these were reminders of a bygone age which would themselves disappear before my childhood was over.

Chapter 12

TALK THE TALK, WALK THE WALK

To survive in the Glasgow street culture, it was essential to pick up the lingo. It was always very important to choose phrases that made you sound as streetwise or as aggressive as the next person. The language used was a great signifier that you were a true product of the Gorbals. This was a completely different form of communication from that used by the middle classes who lived only a few miles away in other parts of the city. For example, if you were threatening to beat someone up, you said, 'Ah'll pelt your melt in,' or, 'Ah'm gonnae kick yir heid in.' A friendly greeting to a male was 'Awright, big man?', 'How's it gaun?' or 'Gaun yirsel, wee man'.

It was important to get your insults right in an aggressive situation. The most useful phrases were 'Get tae f***' or if you wanted to accentuate the insult, you said, 'Get tae f***, ya bampot.' Threatening someone with violence usually involved a quick use of the expressions 'Dae ye want tae go ahead?', 'Fancy a square go?' or if you were feeling really confident and cheeky, 'Ah'll rattle yir baws fur ye.' Sexual innuendo was prominent in Gorbals communication when talking to women you wanted to insult. Vulgar phrases such as 'Any chance o' a ham shank' or 'Can I have ma Nat King Cole tonight?' were used. On entering company, a confident first line was 'Whit's the Hampden roar?' ('What's the score?') If the conversation developed into more aggressive terms, to instigate a fight you could say, 'Dae ye want ma photograph, ya tube?' (If you weren't scared, the best reply to that was, 'No, ye're too ugly.') I often heard women insulting someone by shouting, 'Shut yir big geggie or Ah'll dae it fur ye!' Or to put someone down, they said, 'Ach, yir arse in parsley.'

It is difficult to explain the lingo used in the Gorbals but Stanley Baxter got the nearest to it in his 1960s TV show with the 'Parliamo Glasgow' skits. He had a Glasgow character called Bella Vague who used the words 'thingme' and 'thingmejig'. She would say: 'If you want me thingme . . . ring me.' Baxter also captured the knack local people had for stringing a whole sentence together into one word. For example, describing two pears on a chair thus: 'Errapairapearsonachairoerthere.' This technique was part of the secret of communication in the Gorbals, which sounded completely alien to anyone outside of the area. It was like having your own language within a language:

Thatsnaeborrataw.	It is no bother to do that.
Witastoater.	That's a nice-looking woman.
Geesalightweedoll.	Do you have a light for my cigarette, Miss?
Gonnaepirritinapoke?	Please could you put it in a carrier bag?
Thatauldfoolkintalkthehindlegsaffadonkey.	That pensioner loves to talk.
Watchooterrapolis.	I think we'd better go, I see the local constabulary approaching.
Hawdoanaminuteamdainsumthin.	Please wait for a minute until I finish this.
Aryegonygotaethebrooansignoan?	Are you going to the Social Security office today?

The art of Gorbals conversation was developed by windae-hingin' housewives, who'd get into banter with their neighbours as they watched the goings-on in the street. The opportunities for these chats disappeared when tenement buildings were knocked down and the tower blocks were put up, but boffins have since developed new technology in an attempt to recapture the auld patter. In the 1990s, the Edinburgh University Centre for Speech Technology Research set up computers so that they were able to speak in Gorbals dialect. Not only that, but they were able to speak to each other as if they were having a stairheid argument. Artist David Cotterrell consulted with modern Gorbals residents, who were asked to provide their voices, providing the computers with the ability to speak in Gorbals patter. They recorded the speech patterns of remaining Gorbals people young and old, so the local accent will be preserved.

The Gorbals people have always had their own unique expressions and style of banter. The influence of the language used by groups such as the Irish and Jewish communities in the area ensured that Gorbals speech was unrivalled. In Edinburgh, you can sometimes find it difficult to tell if someone has a Scottish or an English accent. In the Gorbals and other working-class districts of Glasgow, there is never that confusion. Some academics argue that the Glasgow dialect is not just a dialect of the Scottish language but is probably the most common dialect in Scotland, with even outsiders trying to emulate it. When I was at school, the Gorbals dialect was frowned upon and teachers from places like Kelvinside would often tell me that I couldn't speak properly. Nowadays, the children in schools are encouraged to speak in their native tongue instead of being pushed to pick up false posher accents.

There was an old story that summed up the language barrier pretty well. A well-dressed lady with a pan-loaf accent visits a Gorbals tenement. A boy answers the door speaking fluent Glescaranto:

'May I speak to your father, sonny?'

'Oh, my farra went oot when my murra cam in.'

'Then perhaps I could see your mother.'

'Naw, she went oot when ma brurra cam in.'

'Dear me! May I see your other brother?'

'Naw, he went oot when Ah cam in.'

'Isn't there anyone in the house I can talk to?'

'Oh, issiz no ra hoose, missiz, issiz ra cludgie.'

There was a particular tough Gorbals boy I knew who I never saw in a fight. After coming out of approved school aged 17, he realised that he didn't have to have a punch-up with anybody because he had practised the Gorbals lingo to such an extent that he could talk a very good fight. 'Ye don't need tae fight if yir patter is as quick as mine. The secret is tae talk fast and talk loud. That puts anybody aff wantin' a square go wi ye,' he advised. His youthful bravado coupled with fast impertinence made many a so-called hard case extremely wary of him. It was clear then that what you said and how you said it could be far more powerful than any physical action.

To go with the talk you had to have the walk. It was a matter of developing your own hard-man swagger. Everybody's walk was different and mine involved a slight strut and speedy movement of the arms. It took months of practice in front of the mirror and once I had it perfected, I was very proud of it. Having a hard man's walk gave you some kind of social prestige. We were all ruffians in the roughest and toughest area in Glasgow, so it was essential to walk and talk as tough as possible.

Chapter 13

RAT-TRAP

One of the biggest health hazards in the Gorbals was the rats. The various enthusiasts who had built doo huts in the back courts and flew their pigeons in races all over the country complained to us that the rats were eating their prize birds. Some Gorbals pigeon flyers even had small doo huts on their window ledges, tending to them as if they were Burt Lancaster in *Birdman of Alcatraz*. For these men, the pigeons were their lives and symbolised a freedom that they would never have. The pigeon fanciers, many of whom were unemployed and had had a bellyful of living in poverty, knew they were cooped up in the Gorbals with no chance of ever flying away from the place. In a way, we felt sorry for them.

After they complained to us about the vermin destroying their feathered friends, we decided to help them by going on rat-catching expeditions. They were easy to find; dozens of rats scurried about the back courts like wild animals in the jungle. The bigger the rat a boy slaughtered, the more he was admired. Hours were spent hunting them down and there were enough rats, large and small beasts, to keep us occupied all day. I was walking through the back court one morning when I saw what I thought at first was a cat eating a dead pigeon. But on closer inspection, it turned out to be an enormous rat. It quickly scurried off into a deep hole near a midden. A few days later, I saw the same rat again, easily recognisable by its huge size and the large blood-red stain on its back. A gang of us were determined to hunt it down and armed ourselves with large bricks and stones.

We waited for days on end for a sighting but this particular rat was nowhere to be found. We had all but given up; that was until, one afternoon, Albert spotted the rat just outside of an overflowing midden. 'Oh, ya beauty,' he shrieked, 'there he is, there!' And sure enough it was. We picked up a dozen or so bricks and began throwing them but this was a nifty and resourceful creature who managed to scuttle off down the nearest black hole. Albert threw a half-brick full force at the rat. It would have killed it outright, that's if it had hit him. But it missed completely and crashed through the window of a flat as a wee woman was having a quiet cup of tea. She shouted out of her broken window, 'Ya bunch o' f****n' hooligans! Ma boys will kill ye when they get hame.' We ran off but were a bit worried, because her two boys were known as a couple of psychopaths from Thistle Street who had just been cleared of an attempted murder charge at Glasgow's High Court. The papers had alleged they had almost sawn a man's head off because he had drunkenly insulted them. They would be none too happy when they got home to hear from their old mother. We faced a kicking or worse when they found out. We all wanted to keep our heads intact and luckily we knew a young apprentice glazier who owed us a favour. He turned up at the woman's house and quickly put a new window in, saying we were only 'daft wee boys who had been doing a bit o' rat catching'. We pleaded with her not to tell her boys and fortunately she agreed. So the situation was covered up and we escaped a fate not worth thinking about, but it made us even more determined to hunt the rat down.

Some of the boys on the rat-catching jaunts were so scared they tucked the bottoms of their trousers into their socks just in case a rat jumped up their leg. A gang of us managed to hunt down smaller and less agile rats but the big rat still evaded us. One of the boys even managed to collect the dead rats he had caught and carried them around tied to a big stick. I don't know what his neighbours thought, because he often went back to his house for a jam sandwich when his mother was out, still carrying the rats on his stick.

One day, completely out of the blue, Albert spotted the big rat again and began raining stones and bricks down on it. But all of them missed. Then a young fellow appeared from nowhere with an air rifle and shot the rat in the chest. The big rat was badly injured but still managed to limp off into one of his dark holes, leaving a trail of thick red blood behind. The young guy said to us, 'Ah love livin' in the Gorbals. Who needs to go on safari in Africa when ye've got aw this wildlife here? Absolutely magic it is.' It turned out the fellow who had shot the rat was the son of the woman whose window we had smashed. We just thanked our lucky stars that he never found out what had happened, otherwise he would have shot us instead.

Over the years, the Gorbals rats were never out of the newspapers. After a rat bit six-year-old Annette Wilson in 1956 in a condemned Florence Street tenement, her mother said, 'I've lived in this house for fourteen years and the place has been infested with rats all that time. My husband has caught forty rats in the house.' The same year it was reported that four teenagers from Florence Street – John Kinnard, Michael McDonald, William McMann and Dicky Dawson – organised a rat hunt in the area and killed nineteen rodents. One boy told a reporter, 'I hadn't anything to hit them with, but I got three with my feet.'

In 1960, Thomas Ward, fifteen months, and his sister Elizabeth, eight, were rushed to hospital after being bitten by rats while they slept in their Mathieson Street home. A neighbour, Mrs Annie Gill, said, 'When I look out of my window, it's like a scene from the dark ages.' A short time later, ten-month-old Elizabeth Walsh was bitten by a rat in her home in Thistle Street. Her father said it was not unusual for rats to come out and watch him eating his dinner. When a policeman came to the flat, a rat ran across the floor in front of him. The bobby threw his truncheon at the rodent but missed. Interviewed in her rat-infested single end in South Portland Street, Mary Galloway, aged 80, said the rodents didn't bother her: 'It's not the four-legged ones that scare me. It's the two-legged ones I hate. They're more dangerous.'

She had a point. It was not only rats that were being shot during that time; human beings were also victims. One fellow got shot in Crown Street over a gambling debt and another man was shot at the Gorbals Cross toilets in a gangland feud. At that time, it was quite easy to get a gun. There was one gangster type who drank in the Wheatsheaf pub in Crown Street who could get you a revolver if you had the right money.

He even boasted to us that he had sold mass murderer Peter Manuel one of his guns in the past. The guns were usually old army munitions from the war. Some had been brought over by American sailors hoping to make a fast buck by disposing of them in Glasgow.

The Gorbals was hardly Chicago but some of the older men carried guns for protection. When there was gang warfare going on, the bullet was seen as being far mightier than the razor. If anyone did get shot in the Gorbals, which was rarely, it was usually because they had double-crossed someone or, rather ironically, ratted on somebody.

Chapter 14

THE BIG MELTING POT

In the Gorbals of the 1960s, no matter if you were Polish, Jewish, Italian, Irish, Asian or a Highlander, you became a Gorbalian at the end of the day. The Gorbals was one big melting pot where each culture learned about the others' attributes and weaknesses. Being brought up in such a varied cultural environment gave local people the confidence to go anywhere in the world and mix with all sorts of people. Perhaps that's why there are so many Gorbals folk all over the world, including places like the USA, Canada, Australia, New Zealand and South Africa. We were made to be worldwide travellers because we were brought up in an international community.

The Irish made a tremendous impact on the community and had been coming to the area for as long as anyone could recall. They had fled poverty and unemployment in places like Donegal and Cork to make a new beginning in Scotland. The men were hard workers who did not mind getting their hands dirty on one of Glasgow's many building sites. They were big and tough individuals who congregated in the many Gorbals pubs, and several clubs, which had Irish connections. At times, there were so many Irish accents around you could have mistaken the Gorbals for a part of the Emerald Isle.

The Italian community was another one of the largest and most visible groups blazing a trail in the Gorbals. They began to come to Glasgow from impoverished parts of Italy in significant numbers in the 1880s. Many started hot-chestnut barrels on the streets, graduating to ice-cream vans, cafés and fish restaurants all over the Gorbals and the rest of Glasgow. One of the reasons there were so many Italians in the Gorbals was because of what economists call 'the positive multiplier effect'. If someone was doing well in business, they would send for relatives in Italy to join them and they in turn opened up their own businesses. While fish-and-chip shops were unknown in Italy, in Glasgow the Italians made them their speciality. When I walked into Cha Pa Papa's Fish and Chip Shop in Crown Street, Mrs Zeki's in Gorbals Street or Lombardi's Ice Cream Parlour in Rutherglen Road, I often imagined I could well have been in Rome, Milan or Sicily.

Older people recalled early Italian immigrants arriving in the Gorbals with an organ barrel and a monkey to entertain. The kids would dance around the organ grinder, clamouring, 'Hey, Jimmy, can ye no teach the monkey how tae play?' The older Italians boasted that they had walked all the way from places like Naples to the Gorbals. Back in their villages in Italy, they had asked the local priest where the best place was for them to go to seek work in Britain and the priest often recommended Glasgow. They embarked on a three-month journey, walking through Italy and France, catching the boat to Dover and continuing their walk to London then Glasgow. Whole families made the journey that way and there were café owners who even claimed that they had been born on the side of the road during the trip. Some of the Italian children who came to the Gorbals said that they had never seen a pint of milk before because in their villages it had always come out of cows.

The Italians were people accustomed to being in the sun. They often complained that the weather was so bad in Glasgow it was like being in a mine because they never got any sunshine. They were soon nicknamed the 'peely-wally Tallies', because the harsh Glasgow weather and the absence of the sun robbed them of their natural Mediterranean looks, giving them a pale appearance.

Businesses like Lombardi's and the Queen's Café in Victoria Road boasted the best ice cream in Glasgow 'with exciting Continental flavours'. They made a lot of money selling 'pokey hats', which were little ice-cream cones. The story went that the rage for pokey hats had started off years before when an Italian said in his native language to a Glasgow customer, '*Ecco un poco, seniore?*', which meant, 'Try a little, mister,' to which the Glaswegian customer replied, 'Okey pokey,' and thus the pokey hat was born.

Other immigrant groups in the Gorbals, including Highlanders, Asians and Jews, also thrived. Most of the Highland men I knew always said that they hated the English because 'English implantation' had destroyed their communities. In the 1960s, English people were turning up in Highland villages with what the Highlanders called their 'London money' and buying up cottages and farms cheaply. The anti-English feeling came about because of the perception that the incomers knew nothing of the Highland language and culture and expected the locals to bend to their ways and requirements. Some of them arrived in the Gorbals wearing tartan kilts. They were ridiculed by the local diehards, who said they looked like 'big wummen wi beards'. One Stornoway man was known for loudly cursing the English as he staggered through Gorbals Cross. He often uttered an old Highland toast:

> May the best ye have ever seen be the worst ye will ever see,
> May the mouse never leave your grain store with a teardrop in his e'e,
> May ye always stay hale and hearty until ye are old enough tae die,
> May ye still be happy as I always wish ye tae be.
> May the road rise tae meet ye, may the wind always be on yir back,
> The sunshine against yir face, the rain fall upon yir fields,
> And until we meet again may God hold you in the hollow of his hand.

In the 1960s, there were about three or four hundred Asian families living in the Gorbals. Many of the men found steady and well-paid employment with Glasgow Corporation, driving buses or as conductors. According to one of the Asian bus conductors I knew well, Glasgow was celebrated in places like Bombay and Calcutta for offering employment because of its many businesses and industries. Many of his relatives came over to work on the buses. 'We work night and day so that we will have prosperous lives when we get older and perhaps retire back to India. I heard about Glasgow in my village and some of my friends were already here on the buses so I joined them. Living and working in Glasgow is a far better life for me than being in India,' he told me. The Gorbals was their promised land and they sent money to help their families. The poverty of the Gorbals was nothing to what they had experienced in India, Pakistan and Bangladesh.

They had come all that way to break away from deprivation and many of them wasted no time in setting up businesses. They wandered from the Gorbals to places like Bellshill, Motherwell and Bathgate armed with suitcases full of merchandise offering unlimited credit. This appealed to people stuck in the sticks who couldn't afford to go to the big department stores in Glasgow. As a result, many amassed enough money from their suitcases to open up wholesale shops, such as those ranged right along Gorbals Street, supplying a variety of cut-price goods to retailers all over Scotland. Tycoon Yaqub Ali, for example, who arrived in Glasgow in 1953 with a few pounds in his pocket, at first earned his living as a door-to-door peddler. He ended up with the 24-acre Castle Cash and Carry. At its peak, it had a turnover of £100 million and employed 400 people.

In 1967, around a thousand Pakistanis and Indians lived or worked in the Gorbals. By that time they were running twenty wholesale warehouses, fifteen grocery shops and eight clothing factories.

In Crown Street, 'John the Indian', as he was always referred to, built up a thriving grocery shop, working extremely long hours. He opened the shop from very early in the morning until very late at night. If anyone wanted, say, a loaf or a pint of milk, they could go to John's shop at midnight and he would still be open. John had made a lot of money, having such a bustling store in the centre of the Gorbals, and began to show off his prosperity, buying a Mercedes. But some people were taken aback when they saw a little white fellow polishing John's Mercedes. I had never heard anyone being really racist before, but a street-corner man, an old war veteran, looked shocked. He said to me with a shake of his head, 'Wid ye credit it! A white man cleaning a darkie's car! It jist disnae seem right, he must be desperate for the money.'

In the late 1950s, the Muslim community of the Gorbals bought a tenement flat in Oxford Street and converted it into a mosque, the first in the whole of Scotland. It was clear that the Asian community was discriminated against, especially in terms of housing and employment. In 1960, there were reports in the Glasgow newspapers that some schools had been operating a colour bar. Allister McLeod, headmaster of Gorbals Primary School, said, 'There is certainly no discrimination. The chief reaction of the white children is to help the others.' Purnam Singhpall of Logan Street agreed: 'My children are loved by the schoolteachers. They're happy and have the chance of a good education.'

At Catholic schools like St Luke's and St Bonaventure's there were no Asian children at all, although there were many on the streets of the Gorbals. When I asked the local priest about this, he explained: 'To get into a Catholic school you've got to prove you are from a Catholic family. Your parents have to have been baptised and then married in a Catholic church. And you have to be baptised as a Catholic before you can get into this school.'

Jewish people in the Gorbals also made their unique contribution, bringing in their own traditions. There had been a thriving Jewish community in the Gorbals since the nineteenth century and its numbers had been swelled by refugees from Nazi Germany. Several synagogues were dotted around the area, and there were kosher butchers, poulterers and grocery stores selling exotic foods like home-baked bagels and pickled herring. Many Jewish-owned businesses were ultimately very successful, including Fogel's Bakery, Collins' Wholesalers and numerous others. The Jews often said they considered the Gorbals and the rest of Glasgow to be 'a real Jewish city'. The community even had its own newspaper, the *Jewish Echo*, which was published in the Gorbals from 1928 to 1992.

We Gorbals boys sometimes supplemented our pocket money by going to Jewish homes on Saturdays and lighting coal fires. Orthodox Jews believed that it was a sin to light a fire on the Sabbath. Their Sabbath lasted from dusk on Friday till sunset on Saturday. The Jewish families called the people doing tasks for them on the Sabbath '*shabbos goy*' or '*shabbos shiksa*'. Many were proud of their culture and tried to preserve the old traditions and the Yiddish language at home. Some children were banned from speaking English in their homes but were free to do so on the streets.

There were other, smaller immigrant groups in the Gorbals, including Poles, who mostly arrived after the war. We knew a Polish labourer called Stan who had met a local girl at the Barrowland dance hall.

They both had a problem, though: neither of them could speak the other's language. They found out that because of their mutual Catholic upbringing they could converse with each other in Latin. It must have been the first case in the Gorbals of people falling in love after speaking fluent Latin to each other!

There were a lot of 'sausage rolls', as we called Polish people, in the area and pretending to be one of them and not able to speak English was a good way of getting out of trouble. Gangs of boys regularly came up to me in the street asking if I was 'a Billy Dan or an old tin can' or 'a Billy or a Tim', in other words, a Protestant or Catholic. You had to think on your feet. Sometimes I replied, 'I no speak English, I am from Poland,' and then walked off. Despite the fact that the Poles were nearly all Catholic, this seemed to work. If they asked you which football team you supported – another way of finding out what religion you were – the best way to avoid getting beaten up was to tell them you were a Partick Thistle fan. A look of confusion would come over their faces before they let you go.

There was also a small influx of Chinese immigrants, some of whom had opened up a restaurant in Norfolk Street, near Eglinton Street. My mother took me in there when I was aged 12 and I had never seen food like it before. I didn't know what to order but the Chinese waiter suggested that I try chow mein. A big plate arrived at the table for me to sample and I thought it looked like fried grass. The Chinese food went down a storm with the local people and the place was always teeming, especially at the weekends.

The waiter who had served me had a name that we considered unpronounceable and became known as 'Wee Lee'. He must have been on good money because he was always in the bookies putting a bet on. He couldn't speak much English, so my conversation with him was limited to saying hello and giving a nod. Me and the boys noticed that he had a finger missing from one of his hands. We thought he must have been in some sort of fight or accident in Hong Kong. But when we mentioned this to a local gang member, he said, 'Nah, Ah've seen it in the pictures, in *The Terror of The Tongs*. When ye've got a finger missing, it means ye're wan o' the Triads. Wee Lee must be a Chinese

My father in the 1940s. He was always well dressed, even back then. (Courtesy of the author)

My father, Colin, and mother, Betty, before they got married. People used to say they were a smashing-looking couple. (Courtesy of the author)

Crown Street and the corner of Rutherglen Road, 1965. My flat is the top one in the left-hand building. Also pictured are the Wheatsheaf Bar, Coyle's fruit shop and even the bunnet brigade on the corner outside of the bank.

The corner of Cleland Street and Thistle Street. The nearby newsagent sold glasses of ginger for as little as a penny. We played football for hours between the lamp post and the green power box.

Cumberland Street in 1956, looking east from the corner of Lawmoor Street. The Paragon Cinema and St Francis' Church are on the left; Sandyfaulds Street is the background.

A rather untidy back court at 89–91 Hospital Street. How we managed to live and play in such conditions still baffles me.

A typical back court at Crown Street and Thistle Street in 1965.

My father with my brother, Ross, standing at the 48 bus stop in Eglinton Street, waiting to be taken to my granny's in Househillwood, *c*.1965. The picture was taken by me on a cheap Woolworths camera. (Courtesy of the author)

My grandfather, me (right) and Ross in the early '60s, with cheeky Gorbals smiles all round. (Courtesy of the author)

A favourite playtime spot: the childrens' playground at the corner of Cleland Lane and Rutherglen Road. Near by is the back of the Citizens' Theatre.

A back court at Thistle Street and Crown Street, just as I remember it – smoke, prams, kids, dogs, washing lines and middens. A great play area at the time.

The Gorbals boys making a splash down at the Ayrshire seaside. From left to right: Wee Alex, Albert, Ross, my dad and me. (Courtesy of the author)

My father and mother in a Gorbals pub around 1965. At the time he was working as a chef and she as a barmaid. (Courtesy of the author)

The lovely lady workers from the Twomax woollens factory in Rutherglen Road, pictured leaving work in 1955.

Fogell's Jewish baker and grocer, alongside the Clelland Bar in Hospital Street where we watched motors.

The legendary Horseshoe Bar, a big Cumbie haunt, at the corner of Crown Street and Cleland Street. The George Cinema was opposite.

Gorbals Street with the Palace Bingo Hall for the bingo Bellas and the Citizens' Theatre for the luvvies, 1965. The statues on the roof of the bingo hall were taken down in 1979 and now stand in the theatre foyer.

The infamous Gorbals Cross, complete with downstairs gentlemen's toilet. This particular part of the Gorbals is where all the characters congregated – note the windae hingers in the top left corner. Demolition began in 1973 and the toilets went in 1975 – gone but not forgotten.

The 189-foot-high Queen Elizabeth Square dwarfs the old Gorbals of St Francis' Church in the mid-'60s.

gangster.' He laughed before adding 'Ah jist hope that Charlie Chan and his mob arnae tryin' tae muscle in on oor territory.'

For new immigrants, shopping in the Gorbals could be a bewildering experience. For example, if they went into a shop and asked for a loaf, the woman would reply, 'Dae ye want a Mother's Pride, a broon loaf, normal loaf, or a pan loaf, and dae ye want it plain or thick or thin sliced?' Potatoes were also sold in amounts that foreigners found baffling, with the shopkeeper asking, 'Dae ye want a quarter-stone or a hawf a stone of totties?'

Immigration was only one side of the coin, though. While some were beginning new lives in the Gorbals, many Scots were getting out of town fast. Between the two world wars, 600,000 Scottish people emigrated to countries like Canada and Australia. The trend continued during the 1960s.

In 1914, Glasgow was home to over a million people and was the second city of the Empire. It was also the sixth-largest city in the whole of Europe in terms of size. It produced three-quarters of the Empire's ships, half of its locomotives and a great deal of other heavy industrial equipment. Since the 1960s, Glasgow has haemorrhaged population. Today, there are fewer than 700,000 inhabitants. In a 30-year post-war period, Glasgow lost more population than the entire Highlands during the whole century of the Clearances.

During the 1960s, Gorbals people were fed up with being idle with no hope of work. So many took another option, heading for Australia on the £10 assisted passage, to where jobs and sunshine were plentiful. People wrote back to their relatives in the Gorbals saying they had had a fantastic voyage on a luxury liner on the way to Australia. They had rubbed shoulders with real toffs who had paid a comparative fortune to be on the same voyage. But the general feeling at the time was that they might have got what they wanted, but they'd lost what they had. One woman from Gorbals Street wrote back to her former neighbours extolling the wonderful attributes of Australia and said that after the luxury cruise the family were housed in buildings with tin roofs. She said that when the heavy rain hit the building, they all imagined that rats were tap-dancing on the roof.

Some people left the Gorbals for a new life in Australia only to reappear again a few months later claiming it had been too hot for them and they missed the Glasgow weather. One young guy, a joiner called Tam from Florence Street, had a big party on the eve of his departure for Australia with his new wife. He made a long, emotional speech, maintaining that Scotland was the past and Australia was the future. We all wished him well and there was a big turnout to give him a good send-off. So I was a bit taken aback to bump into him a few weeks later at Gorbals Cross. He looked a bit embarrassed as he said, 'Ah took wan look at Australia and Ah missed ma ma and da and aw ma pals. Besides, the beer wis pish. Never again! Ye cannae beat being hame wi yir ain people.'

In 1966, the *Evening Citizen* reported that after ten months in South Africa, Patrick and Doreen Donnelly became so homesick for the Gorbals that they stowed away, along with their baby son, Pat, on a liner bound for Southampton. When they arrived they were fined £20 each. Patrick's mother, Mrs Mary Donnelly, told the couple, 'You live in the most wonderful place in the world. I'll never know why you left it.'

Chapter 15

SHOPPING

E ven by the 1960s, there were no large supermarkets in the Gorbals, but punters did have access to a multitude of shops that offered cut-price food. There was a shop in Ballater Street known as 'the Shanny' which attracted hundreds of people every week. The queues often stretched several hundred yards up the road. The Shanny offered cheap bakery items – bread, cakes and biscuits which were misshapen or near their sell-by date. Families in the Gorbals lived quite well because of the Shanny and the goods rarely tasted musty or stale.

The local shops were quite unlike today's impersonal supermarkets. The women behind the counter usually knew what your name was and asked about your family, saying things like, 'How's your mammy getting on? Is yir faither still workin' away? How's yir wee brother doin', awright?' When my mother was short of money she would write out a note and dispatch me to the grocer's with it. She was too embarrassed to go herself because it was asking for 'tick' – credit. You could get food on loan until the end of the week, when the bill would be settled on pay day. The shopkeeper usually complied with the request. When I took the note down to the grocer's, the woman behind the counter would give a knowing nod and mark my purchases in her tick book. Most businesses in the Gorbals operated this system.

But that changed when a new owner took over. He declared: 'All that tick business is no good to me. From now on, we are only dealing in straight cash.' He was from outside of the area and had run a similar shop in Bearsden, where tick was unheard of. We saw him as a bit of a tightwad who took sadistic pleasure in having the power to veto people's requests for tick. To add insult to injury, he even put a broken clock on the wall behind the shop counter with a sign attached to it declaring 'No tick here'. But his custom quickly fell off as people avoided the shop saying, 'That guy is a real Scrooge, but he'll learn, don't worry aboot that.' He did indeed learn: the boycott of his shop accelerated and a few of the boys caved in his shop windows one night in a brazen act of vengeance. After a short time, the offending, no-ticking clock was removed and the shop was back to offering credit again. Tick was an essential part of trading in the Gorbals and if a shop didn't provide credit, it faced the humiliation of bankruptcy.

The local Co-op was one of the favourite places to go for the messages. It operated a sort of bonus system. Some people lived by the Co-op and even died with the Co-op. If folk bought their food or clothes there, or even had a family funeral with them, they got a Co-op dividend, which was given out once every quarter. The dividend provided clothes or food and was seen as a nice perk. The workers there were more polite than the average shopkeeper. They had been on training courses advising them that the paying customer was always right and never to talk about religion or politics. They were also instructed never to gossip with one customer about another. The Gorbals was full of gossips and it must have been frustrating for them to encounter an employee who had been on a Co-op training course. A neighbour complained to me, 'They Co-op people are no allowed tae gossip but, for goodness sake, gossip and a good blether is whit makes life mair interesting. If ye cannae talk aboot people behind their backs, whit's the point o' going in there?'

At the weekends, families went to the Barras. The Barrowland market was, and still is, a centre for the hustle and bustle of Glaswegian life. It was held on Saturdays and Sundays, and had hundreds of stalls, both indoors and outdoors, selling everything imaginable. It was very cheap compared with the high-street stores and it was said that a lot of the merchandise was obtained by dodgy means.

Having a bag of whelks or winkles was part of the experience. We were provided with a pin to extract the little sea creatures from their shells. Some of the boys detested them and found the look of the worm-like creatures pretty disgusting, saying, 'That looks like a big snotter that's jist come oot yir nose.' But I wasn't put off – I considered them to be a delicacy.

The Barras was always good for the banter and people flocked there to hear the latest line in patter. There was a sharp fellow from Nicholson Street who was always there wheeling and dealing. We nicknamed him 'Bobby the Barras'. He had a very sharp tongue and was usually in various pubs in the Gorbals doing some dodgy deal involving knocked-off gear – he was actually quite open about it. As far as we knew, he was never arrested, so we presumed he must have had a few of the local constabulary on his payroll.

Bobby had an incredible knack for the patter. His chat was very fast, with all the words merging into one: Howsitgaunboysawright? ('How are you, boys, I hope all's well?') Isyirmammylookinfuranewiron? Andisyirauldmanwantanewovercoatfurthewinter? ('Does your mother want an iron and your father a new coat for the winter?') Tellemtaecomeanseemeatthebarrasandahlldaethemaguiddeal. ('Tell them to come and see me at the Barras and I'll give them a good deal.')

At the Barras, Bobby slowed the sales patter down so that the tourists from outside of Glasgow could understand him. 'I don't want a fiver, three pound, two pound or wan pound. No even ten bob, five bob or four bob . . . give me two bob a pair.' During the resultant frenzy, hundreds of hands flew up in the air. Bobby's patter went on and on like a verbal machine gun: 'If ye cannae see a bargain here then God help ye . . . No, sir, they are not stolen, it's jist that Ah hivnae paid for them yet . . . Dae ye want yir change missus or a share of the business? . . . The polis don't even know this is stolen yet . . . That's hawf yir retail price, hawf yir wholesale price and even hawf yir hawf-price . . .'

Bobby and the rest of the traders worked on low profit margins, with a high turnover and a humour that always ensured big sales. Many of the Barras auctioneers could have been professional comedians. When drunks appeared in the crowd to heckle, they dealt with them swiftly, ensuring they did not come back again.

Most of the time, Bobby simply admitted that the goods had been knocked off; it made them sell quicker. He'd shout, 'These immaculate tea sets were headin' fur the House o' Fraser until we got our hauns on them. The difference is that the House o' Fraser will sting ye for a hundred pounds, but gie me a fiver an' ye can hiv a tea set better than the ones at Buckingham Palace.' One time, Bobby came into possession of hundreds of packets of washing powder and he shouted, 'Get yir top-class washing powder here cause everybody needs to dae a washin', unless ye're a clatty bastard.'

The Barras was awash with colourful characters. Just along the street from Bobby's pitch, a big African man sold exotic so-called snake oil, which he claimed was 'from the dark jungles where no white man has set foot'. Traders also made money capitalising on the city's Catholic–Protestant divide and sectarian bigotry. The stalls that dealt in Old Firm gear sold scarves, flags, hats, posters and T-shirts proclaiming such bold messages as 'Rangers, Kings of Scotland' or 'Celtic, '67 European Champions'. They even sold tiny baby suits bearing either Celtic or Rangers designs and motifs.

The Barras was also a splendid place for a sing-song on a Saturday or Sunday morning in the various pubs dotted around the area. It was there you heard the best and worst of Glaswegian crooning. The turns ranged from people who sounded as if they had stepped straight off a West End stage to the diabolical singing of out-of-tune drunks.

After work, Bobby and his trader friends went to the Saracen Head, known as the Sarrie Heid (although regulars with hangovers called it the Sorry Heid). Inside there were always dozens of eccentric personalities. The pub was built in 1754 and it was where, in the old days, the stagecoach used to depart from Glasgow for its 12-day journey to London. Wordsworth and Burns, as well as Johnson and Boswell, had visited the inn at a time when it was the rendezvous of the town's elite.

Behind the bar was a huge figure called Big Angus. All the locals drank what they called champagne cider – 'a shammie' – and White Tornado wine. Bobby and the other regulars would walk up to the bar and say to Big Angus, 'Ah'll hiv a shammie and a White Tornado.' The combination of the two drinks gave the consumer the feeling that all was well with the world and that he had been sedated against life's everyday problems. Having a shammie and a White Tornado affected people differently. Some became uproariously funny, with brilliant patter flowing. The White Tornado made others maudlin about their lives and inevitably a sentimental song, perhaps an Al Jolson number, would float out the door of the smoky, atmospheric pub. Old Scottish songs were popular and as we passed by one day the whole pub was singing 'We're No Awa tae Bide Awa', which had a nostalgic ring to it:

As Ah wis walkin' doon the Gallowgate,
I met wi Johnny Scobie,
I says 'Man, will ye hae a hawf?'
He says, 'Man, that's ma hoabbie!'

For we're no awa tae bide awa,
We're no awa tae leave ye,
For we're no awa tae bide awa,
We'll aye come back an' see ye.

The inevitable arguments and fights broke out at the Sarrie Heid but Angus was no stranger to sorting out skirmishes and throwing people out for behaving obnoxiously. He was said to be an ex-policeman and he dealt with minor incidents by poking his large finger, which looked like a gun, into the culprit's face, saying, 'You've been causin' trouble here, pal – time ye disappeared.' When he had to throw people out, Angus was a formidable sight as, like in the cowboy films, he gave them the bum's rush through the door.

Angus initially looked like a rough and ready character but when you actually got into conversation with him, he was a fairly respectable, worldly-wise man. He made an extremely good living over the years and had a nice house in a better part of the city, which must have seemed like a million miles from the goings-on at the Sarrie Heid. He came out of the pub one day and we asked him, 'How's it gaun, Angus?' He gave a shrug of his shoulders and said, 'People are always criticising me fur selling these people drink, because nobody else in Glesga will even entertain them. But Ah can put up wi aw the drunks, heidcases and eejits, and the criticism. Sure, Ah'm crying aw the way tae the bank.'

Just across the Clyde from the Gorbals was Paddy's Market. This was a completely different world from the Barras. Paddy's Market was where the poorest of the poor went to buy things that most of us would be too embarrassed to throw away.

The overriding impression was of poverty. Hundreds of people stood in the wind and rain selling what looked like piles of rags and other items such as old tattered books, shoes and records. There were numerous stalls inside, selling everything from used clothes, old records, books and furniture to bashed cans of tinned food to cheap tobacco. Paddy's Market also had a dingy restaurant at the back, which sold its legendary mince and tatties and boiled ham ribs with cabbage and potatoes. To the regular customers, it was the tastiest food in the whole of Glasgow.

I knew of people who went shopping at Paddy's Market in disguise, covering themselves with hats and scarves, because it was seen by some as a comedown to be spotted shopping there. The biggest put-down at school was for someone to say to you, 'Hey, where did ye get that jumper? Paddy's Market?' Or, alternatively, you could be taunted by other kids shouting at you, 'I saw your maw at Paddy's Market yesterday!' Catering exclusively for the very poor, the place served a true social function, and it still does to this day.

Just across the road is the Glasgow Green. We often wandered over there to go into the People's Palace museum, with its large, exotic greenhouse. Another fascinating sight was the nearby old Templeton's Carpet Factory. It looked really incongruous because it was an almost exact copy of the Doge's Palace in Venice. On hot summer days, whole families headed to the Glasgow Green, lounging about in the warm sunshine and throwing stale bread to the swans in the duck pond. The children were treated to pokey hats from the ice-cream van. To the kids, it was like going on holiday. People who could not afford to go away joked about it; when asked where they were going they replied, 'Hameilldaeme' (home'll do me). A baffled look came over many people's faces before they twigged that Hameilldaeme wasn't some exotic destination.

Wandering about the Glasgow Green as boys, we realised that although Edinburgh was thought of as the cultural capital of Scotland, Glasgow had its own unique buildings and way of life which other cities in Britain could not match. One street philosopher standing outside of the People's Palace gave us his unique, if somewhat biased, view: 'People are always comparing Glasgow wi Edinburgh, but there's nae comparison. Edinburgh might hiv a big castle but Glasgow has got the people, the real people. We arra people!'

Chapter 16

THE BUNNET BRIGADE

The older Gorbals men – the 'bunnet brigade', as we called them – stood every night at the Bank Close on the corner of Crown Street and Rutherglen Road with Peter the paper man as he sold the *Daily Record*. The conversations were fascinating for a young person such as myself. Surely, if there was a University of Life, this was one of the top colleges. There were hundreds of different topics, night after night, ranging from football to politics, economics to history, religion to crime, to what was the best tip for the next day at Ayr or Aintree.

Throughout my childhood, these older men stood with Peter philosophising. The noticeable thing was they all looked the same with their tartan bunnets on. It was a Gorbals tradition but why the older men wore these flat caps baffled me. It was the '60s and their dress code had not changed since the '30s. One day I was passing through Gorbals Cross when it all fell into place. There was a bunnet shop with a sign outside proclaiming: 'If you want to get ahead . . . get a hat!' I just burst out laughing thinking of all those old characters putting the world to rights with their tartan bunnets on.

We often joked that Peter was a bit like Dracula. No one knew what he did in the daytime or where he was during those hours of sunlight, but he always appeared when it was dark. To be fair, Peter was better looking than Dracula; but his wit and perception could be just as deadly. Every now and again, he would comment on a particular individual who had just bought the *Record* from him. After one guy walked away, he'd say, 'That fella would steal the sugar oot o' his granny's tea.' Or commenting on some fellow's luck in getting off with a criminal charge, he'd quip, 'If that guy fell into the Clyde, he'd come oot wi a fish supper!' It was said that Peter knew everybody and everything in the Gorbals.

One Friday night after the pubs had come out, a guy in his 30s, quite drunk, came over to talk to Peter, myself and the bunnet brigade. He was boasting that he had just beaten up one of the Cumbie gang outside of the Clelland Bar. He was full of bravado due to the drink and was celebrating his so-called great victory. I knew the fellow he was talking about and was surprised – that's to put it mildly – that this guy had hammered him. Afterwards, when he'd left to boast to more people, Peter said, 'Aye, he's bragging aboot it the noo. But he'll no be bragging aboot it the morrow morning when the big man makes his comeback.' He was right enough and the bunnet brigade nodded in agreement. The big man turned up at the guy's door the next day with a heavy team of his pals. The drink had worn off and the boaster was reduced to giving a grovelling apology. 'I had that f****n' bampot shaking like a leaf after he apologised,' said the big man to Peter a few nights later. 'It's lucky he didnae get done in. Ah used tae go oot wi his sister, that's the reason he's still breathin'. Otherwise he'd be proppin' up the daisies. Ah gave him a good punch in the mooth, because it wis that mooth that landed him in aw this trouble,' he added with a touch of nonchalance.

Some of the bunnet brigade prided themselves on a lifetime in the shipyards at places like Harland & Wolff and Fairfield's. They described the shipyard atmosphere in graphic detail: 'There wis this continual dull thudding, night and day, as the riveters and their apprentices went tae work. They worked through aw sorts o' atrocious weather, including rain, sleet and snow. Even then the majority o' the men never even thought aboot taking a day aff.'

Two of the men had worked together for years in the shipyards as riveters and took great pride in their profession. One was a 'putter-in' and the other a 'hawder-oan'. Together, they reckoned, they had been the fastest riveters on the Clyde. The hawder-oan said, 'There's a knack to oor job. After a while, ye build up yir speed and can put in the rivets wi perfection. When the ship was finished, the workers turned up at the launch cheering and waving flags. They were stupid bastards, though. Because once a ship was launched, we usually ended up getting laid aff.'

A favourite subject with the older men was their tales of survival during the Great Depression of the 1930s. They had joined the hunger marches and rioted against the Household Means Test, which had been brought in by the Government to save money. It punished the poorest, and some Gorbals families were forced to live on a couple of shillings a week. If they were turned down for money, they appealed to a Means Test tribunal and still might get nothing. So many turned to crime or wheeling and dealing to survive.

'The Means Test inspectors came to yir hoose and told ye tae sell yir furniture, and they even looked to see how much coal ye hid in yir bunker,' one old fellow told us. 'There were three million of us oot of work and some people called those withoot a job lazy bastards. But how can ye hiv three million lazy bastards? There wis nae chance of a job. The geniuses of the world were all standing aboot on street corners doing nothing. Skilled men – engineers, miners, shipbuilders, builders – aw wi nae prospect o' work. So we ended up rioting and stealing tae make our point. The only people wi money then wis the polis. They were aw on good money and overtime. It wis nae wonder we turned tae crime, it wis the only way tae survive.'

Although I was only a kid with an inadequate education and a limited knowledge of the world, I understood then that the crime and the hatred of the police in the Gorbals had historical roots. These old men were living proof of that. 'We were aw starvin' and the polis were makin' barrowloads o' money, so we hid to riot, we hid to show them that we hid strength,' said the old guy. He was so bitter that if he saw a policeman walking down the street, he'd shout obscenities at him, but he never got arrested, probably because of his age.

The Second World War had certainly made its mark on the bunnet brigade. They told tales about the bombing raids on Glasgow and even about spending time in prisoner-of-war camps. There were quite a few old air-raid shelters dotted around the place. They were great places to play in and every boy imagined he was fighting the Germans. A lot of people had even kept their gas masks from the war. My granny still had hers neatly packed in her broom cupboard and she never threw it away. I asked her if she kept it for sentimental reasons but she just shrugged and said, 'Naw, Ah'm no sentimental. Ah'm keeping the gas mask jist in case another war breaks oot. It could come in handy. I've been through two world wars and after the first one we thought it would never happen again but then Adolf Hitler started tae dae his stuff.'

There was much talk about the Great Blitz, which had hit Clydebank. One man said he had been an ambulance driver at the time, and when he drove towards Clydebank, the bombing was so fierce it 'looked like the Blackpool Illuminations'. He told us: 'You should hiv seen it. The Germans dropped incendiary bombs first to light up the place and then the bombers flew in dropping their big bombs. It was wave after wave of fire and bombs but the strange thing is we never really felt scared. It was all that adrenalin pumping through yir body. But afterwards Ah sat doon in the pub and had a double whisky tae stop me from shaking. It was then ye realised that the shock had hit ye.'

The Japanese were particularly hated by the old-timers, as some of the men had suffered at their hands. These men depicted them as 'cruel slitty-eyed bastards' who had starved them and had even 'cut the heads aff oor pals'. One old Scots Guard even had a gory story, which he seemed to relish, about the enemy playing football with his decapitated pal's head. 'They cut his heid aff and kicked it aw ower the place, jist because he wis a bit cheeky tae them,' he said. Whether it was true or not, I do not know, but he certainly captured everyone's attention. There was also talk of men being hit with bamboo canes, forced marches, jungle warfare and torture, including being made to stand in the blazing heat. One of the bunnet brigade, Tommy, who was in his late 60s, had been a Japanese prisoner of war and often regaled his street-corner audience with his tales. He harked on, 'They starved ye until ye were only skin and bone and then they beat ye intae a pulp.' He always ended his lectures with the words: 'Never trust a Jap.' I would have heeded Tommy's advice but even though the Gorbals was full of all sorts of immigrants, I never bumped into anyone Japanese.

There were other stories of heartache. Men came back from the war with completely changed personalities. Their children did not know who their fathers were when they came back to the Gorbals because they had been away so long. There were divorces and separations because Daddy had become a stranger and perhaps had hit the bottle too much. Whisky might have taken away some of the symptoms of depression but it did not take away the root of the problem. As Tommy explained: 'It wis an absolute disgrace, so it wis. There wis naebody tae help they poor men after they walked oot the gates of the camps. They were abandoned and had nae chance o' fending for themselves.'

Widows of the war abounded among the older women of the Gorbals. Many of them still had keepsakes or souvenirs of their men which had been sent to them by the army. It might be a comb, a pen, fingernails or even a lock of hair. Some of them were sceptical about whether what they had received had actually belonged to their late husband. Others refused to talk about it; they blotted out the memories, saying, 'Whit's the point o' talkin' aboot it? There's nae point, it's happened and that's that. If ye talk aboot it, it only gets worse. Ye cannae turn the clock back.'

Everybody had a story about the war but the problem was that, to us kids at least, a good number of the tales sounded the same. It could be a bit tedious when we were talking about something and one of the bunnet brigade commented, 'You're lucky, son. It wis a lot worse than that during the war.' However, talking to the old boys night after night, I realised that I was learning more about history than I ever would at school, so I made sure I listened to all the stories.

There was an unusual tale that stood out. An old, grey-haired woman called Mrs McGraw said to us one day, 'Every auld geezer's got a war story but if you come to ma hoose, I'll show you somethin' that beats them aw!' The next day, me and the boys went to her single end. She gave us a warm welcome, invited us inside and said, 'Ah've been dying tae show somebody this for years but Ah've hid tae keep it a secret until noo.' She opened up a small cupboard and inside there was what appeared to be part of a large, dusty white blanket. We were disappointed, saying, 'Is that aw ye've got tae show us, missus, an auld mingin' blanket?' But she explained: 'No, that's no a blanket, son. It's Rudolf Hess's parachute – Hitler's right-hand man.'

She told us a fascinating tale. On 10 May 1941, local people saw a parachutist float into a meadow at Floors Farm in Eaglesham, near

Glasgow. They ran out to find a crashed, burning Messerschmitt and a lightly injured German officer calling himself Captain Albert Horn. But it turned out to be Rudolf Hess. He said he was on a solo mission to end the war and had hoped to land at the nearby estate of the Duke of Hamilton. Hess was under the mistaken impression that Hamilton was the leader of a British peace party. Mrs McGraw told us: 'My man wis there in Eaglesham that night when Hess wis arrested. Wullie thought it wid be a good idea tae take part o' his parachute as a souvenir. Ah've kept it secret for aw these years in case the authorities found oot. But whit can they dae noo? Wullie's gone and Ah'm too old tae send tae jail. Anyway, whit wid they want wi the parachute noo?' After his capture, the Nazi leader was imprisoned by the British for the rest of the war, after which he spent the rest of his life in Berlin's Spandau Prison. Had he known, I suppose it would have dumbfounded him how his parachute ended up in a single end in the Gorbals. It hardly surprised us; Alex summed it up by saying that a lot of things that went missing in Scotland ended up in some house in the Gorbals.

After seeing the parachute, a team of us, Alex, Albert, Chris and me, cycled all the way from the Gorbals to Eaglesham, which took us about two hours. We put up a tent in the woods beside the river, lit a roaring fire and camped out for the night. Underneath the midnight stars, we tried to imagine what Hess had felt when he landed in Scotland. It was crazy – we Gorbals boys attempting to retrace Rudolf Hess's footsteps.

One member of the bunnet brigade, Benny, was always telling us tales about how he had been round the world with the merchant navy. Some of his stories were so incredible that he should have written a book about his adventures. Travel had given him a wisdom, an experience and an inner confidence that made him stand out from the other men. They could not match his storytelling technique.

One night, Benny handed me a scrap of paper and said, 'Read this, son. I saw this on my travels and the words have always given me confidence, through thick and thin. Take them home and, like me, memorise them. Try tae remember that even if ye are a wee diddy in the Gorbals, ye can still go out and conquer the world. Success starts with the mind. Naebody knows where confidence comes fae or where it goes. But if ye remember these words, it will always come back tae ye.'

I got back to Crown Street and unfolded the scrap of paper in front of a roaring coal fire. It was clear the words would instantly fill anyone with confidence:

> If you think you are beaten, you are;
> If you think you dare not, you don't.
> If you like to win, but think you can't,
> It's almost a cinch that you won't,
> If you think you will lose, you're lost,
> For out in the world we find,
> Success begins with a fellow's will,
> It's all in the state of mind.
>
> If you think you are outclassed, you are,
> You've got to think high to rise,
> You've got to be sure of yourself before
> You can ever win a prize.
> Life's battles don't always go
> To the stronger or faster man;
> But soon or late the man who wins
> Is the one who thinks he can.

The funniest raconteur of the bunnet brigade, however, was Big Archie, who really fancied himself as a comedian. He was a tall guy in his late 30s who never stopped telling jokes and stories. He also thought of himself as a bit of a ladies' man and often stood outside of the Twomax woollens factory in Rutherglen Road trying to chat up the girls. He said to me one day, 'That factory employs some o' the most beautiful women in the world. Staun ootside wan mornin' and watch them going in, they're aw gorgeous creatures.' I did, and he was right: the factory seemed to have a knack for hiring extremely pretty women.

Archie, a divorcee, was fond of a drink but because he was out of work most of the time and short of money, he only ventured into pubs now and again. So most nights he stood on the corner trying to be as funny as possible. We said he should have been on the stage, to which he'd reply, 'Aye, it leaves twelve o'clock!' Then he'd launch into his stories.

His Auntie Bella was one of his favourite comic characters. Whether she actually existed was another matter. 'Ma Auntie Bella wis comin' hame fae the Palace Bingo when somethin' came over her . . . it wis a double-decker bus. She loast baith her legs but she wis fittit wi a perra widdin wans by the National Service. She managed fine wi them until wan night her chip pan caught fire and set the hoose ablaze. The hoose wisnae badly damaged but ma auntie wis burned tae the ground.'

Archie had a knack for telling really funny stories about the wild young gang members that roamed about the Gorbals. 'Wee Rab and Tam bumped into each other at Gorbals Cross. Rab said, "Ah hid a rare night last Friday. Ah wrecked a fish an' chip shoap, pulled oot ma razor and slashed the owner, sprayed paint aw ower the place and grabbed two fivers fae the till. Then the polis lifted me."

'Tam asked him, "How did ye get on in court, Rab?"

'"Awright," replied Rab, "ma Mammy spoke up fur me. She told the Sheriff that Ah was well brought up and had a lovely nature and never did nothin' wrang."

'"Oh, that's smashin'," said Tam. "Did ye get aff, then?"

'"Aye, sure," said Rab, "all that happened wis ma maw got nine months for perjury."'

Another of his favourite stories involved people being moved from their old tenements to the new tower blocks.

'Ma pal Wullie got moved tae wan o' they new high flats. But he felt like a prisoner 12 storeys up and he got dead depressed, so he thought he'd throw himself intae the Clyde. He was stood on the parapet o' the Suspension Bridge when a big polis shouted, "Don't jump!"

'"Why not?" Wullie asked.

'"Because if ye jump intae the water, Ah'll have to jump in tae try and rescue ye and Ah cannae swim so Ah'll end up drowning. If Ah drown, ma wife and seven weans will have naebody tae support them."

'Wullie replied, "Ach, Ah'm gonnae jump anyway."

'"No, hang on a minute," shouted the big polis, "Ah've got a better idea. The water in the Clyde's freezin' cold and manky. Why don't you go back tae yir nice wee bachelor flat and hang yourself in comfort?"'

That story of Archie's reminded me of the time I saw a man dive headlong into the Clyde one dark winter's night. When I looked around to see if there was anyone chasing him, there wasn't. He had decided to do himself in by drowning in the Clyde. I don't know what happened to him, he just sort of floated away. But there was a lifeguard who had a rowing boat in the Glasgow Green stretch of the Clyde and he told us it was his job to go up and down the Clyde looking for bodies. Every weekend he usually found a couple of them. It was a strange job but I suppose somebody had to do it.

One night, Big Archie did not turn up and Peter said he had heard he was in a bad way in hospital. Peter explained, 'He had a win at the bookies and began tae drink a lot o' whisky. He joined a couple o' the Cumbie who were wi their birds in the Wheatsheaf bar. Next minute, Archie began telling dirty jokes tae them. They didnae like it wan bit and Archie ended up getting slashed. He thought he wid hiv them in stitches but Archie's the wan in stitches noo. He should hiv known no tae tell dirty jokes in front o' women. He made a big mistake but, mind you, we aw make mistakes – that's why they put rubbers on top o' pencils.'

Chapter 17

DODGY CHARACTERS AND POOR SOULS

Even leaving aside Peter the paper man and the bunnet brigade, there were hundreds of diverse characters in the area, some of them pretty dubious. Gorbals Cross was where we met most of them. It was a hustle-and-bustle place where a myriad of humanity converged. For example, there was Harry the pickpocket, who travelled all over Britain relieving people of their wallets. The amazing thing was he never got caught because his hands were so fast and his fingers so nimble. Alex knew him well, even claiming at one point that Harry was his long-lost uncle. On one particular winter Saturday night at a bustling Gorbals Cross, Alex told 'Uncle Harry' he was short of money. Harry was a short guy with big thick glasses, wearing a dark suit. There was a glint in his eye that said, 'Ah'm takin' on the world, jist watch me go.' He walked over to one fellow and began asking him for directions to somewhere or other. Astoundingly, we didn't even see his hands move, they were that quick. The next minute, he came over with his hands concealing a wallet and gave Alex a couple of pounds. 'There ye go, boys,' he said, 'another mug, another day.' We often asked Harry if he could train us to become professional pickpockets but he just laughed and said, 'Who d'ye think I am? Fagin fae f****n' *Oliver Twist*? Pickin' pockets is a natural gift, it's self-taught. You've either got it or you've no.'

Unsavoury characters often stopped for a bit of patter and tried to impress us with the various criminal skills they had. We got into one such conversation with a big guy we had nicknamed 'Scarface'. He pointed to one scar on his face and said, 'Bridgeton, 1962.' He pointed to another saying, 'Govan, 1963,' and it went on and on. This man had so many scars his face and hands were like a diary of his fighting exploits. Because he had so many, we were not entirely convinced that he was a real hard man; more of a mug that was getting slashed all the time, we thought. But a leading member of the Cumbie gang said that this guy was no bampot. He had been in hundreds of fights involving open razors and knives, and was more often a winner than a loser. 'Put it this way,' said the Cumbie guy, 'if he wis a mug, he wid be well deid by noo. Forget aboot the scars, that guy really knows how tae handle himself. Ah widnae even consider tackling him. That guy is hawf mad.'

One afternoon at Gorbals Cross, Scarface pulled out an open razor, showing us how to handle it without cutting ourselves. He then proceeded to demonstrate the best way to slash someone in one swift motion. 'Try no tae go for the throat,' he advised. 'Ye could end up on a murder charge, jist go fur the face if ye can.' I don't think there was anywhere else in the world that you would have got free slashing lessons. He looked at us in all seriousness and said, 'Ah've always thought ye could make a great book called *Where Did You Get That Scar?* wi photos o' lots o' people and the stories behind their scars.'

I thought perhaps this might be a good idea. The Gorbals was full of people bearing scars and it was an accepted part of life. Most guys showed off their scars rather proudly and recounted the hair-raising stories behind them. Having a 'Mars Bar', as we called it, could be a form of social prestige. Like a tattoo, it made an individual stand out in the crowd. A few years later, when I was 15, another boy tried to slash me with an open razor but I was too fast and managed to duck and run before the blade came near my face. I personally thought that a scar was the emblem of a loser and I could certainly live without one.

Another fellow we met, called Danny, gave us shoplifting lectures. He demonstrated how to make things disappear faster than Harry Houdini. 'The trick is, boys', he explained, 'no tae look too conspicuous. Always be anonymous. When ye merge intae the background, the world is your oyster.' He was also an expert at conning people. Danny would knock on someone's door in a posh area and tell the owners that he had a parcel for next door but there was no one in; could they take it in as a favour? They always agreed but he would then explain that it was cash on delivery. The neighbour was tricked into giving him a few pounds. When their next-door neighbours arrived back, they would be surprised to hear that there was a mystery parcel for them which had been paid for. When they opened it up, it would be full of coal ashes.

Another of his ploys was to go to a house pretending to be a builder and then tell the owners, usually elderly, that they had slates falling from their roof. They could end up 'killing some poor soul'. The gullible, of whom there were many, then handed over 'a right few quid', as he put it. A few years after I first bumped into him, I went to the Sheriff Court to see one of the Gorbals boys being tried. By coincidence, Danny was up as well. He defended himself with great eloquence and on being found guilty he even congratulated the procurator fiscal on what a great job he had done.

Like a typical con man, he addressed the court in a posher accent than he usually used, and the sheriff was amused and entertained by him. The procurator fiscal read out a string of his previous convictions over a 30-year period, 66 in all. Because of this, we were expecting him to be remitted to the High Court for a long sentence. But he had put the sheriff in such a good mood, he only got two years. Before being led away, he said to the sheriff, 'Thank you, m'lud. It is always nice doing business with a true gentleman.'

A similar Gorbals con man was 'Bandit' Rooney. There were hundreds of stories about him. The most popular one was that he bought cheap imitation rings and then went up to the city centre where the expensive jewellers were. He would spot a young couple, much in love, gazing in the windows at the engagement rings, which they could barely afford. The Bandit would then approach them with his rings and offer them at a 'knock-down price'. As far as we knew, he made a fair living out of doing this.

At any one time, the Gorbals was full of mad individuals and mad situations. I often thought that most of the people who were locked up for being mad were probably far saner than some of the people I met daily. There was a really crazy fellow nicknamed 'Celtic Joe' who wandered round the streets wearing a green Celtic top. It was a shame, as he should really have been institutionalised. He talked in an aggressive, high-pitched voice and even joked about his condition, saying, 'Ah'm no mental, Ah've got a certificate fae the doctors to say Ah'm no.'

Some of the boys thought it was a great source of entertainment to wind Joe up. So the conversation would start off seemingly normally enough, with Joe commenting on his favourite Celtic heroes. 'See that Jimmy Johnstone, he's a magic wee man. That Ronnie Simpson his got tae be the best goalkeeper in the world . . .' A couple of the boys would then walk across the road, start giving him the V-sign and shout, 'Rangers! Rangers!' Joe's face turned a bright red and his eyes bulged. The next minute, a chase began, with Joe running after the boys shouting, 'Ah'll kill ye, ya Orange bastards!' Sometimes the chase lasted for up to an hour. Joe pursued the boys up and down closes, through back courts, over walls and he even looked in middens where he thought they might be hiding. The boys were too cunning and streetwise to get caught. But they believed that if Joe did catch them, they faced a beating or even being murdered. One of the funniest and most bizarre sights was watching Joe chasing a boy one day who had jumped onto the back of a speeding bin wagon to escape his clutches. Joe was waving his fists and shouting, 'Ah'm gonnae cut your heid aff!'

After the chase had ended, if we bumped into Joe the next day, it was as if nothing had happened, for he had no short-term memory and could never remember what had happened the day before. Much to our dismay, Joe suddenly disappeared from the streets of the Gorbals. The rumour was that he had come to an untimely end, having had a heart attack during one of his chases.

Some were even worse off than Joe, living on the fringes of society, dealt a cruel blow by ill luck and madness. Billy the boxer was one example. He was a well-known pugilist in his day but he had taken so many batterings over the years it had left him brain-damaged. By his late 30s, he was like a sad pantomime character, punching holes in the fresh air as he walked along the street. You don't see many punch-drunk men now but then, because of a lack of medical attention and understanding, they were not uncommon. Billy would suddenly stop in the street and throw a right hook, then a left at some imaginary opponent in an equally imaginary title fight. 'Ah'll take ye, ya bastard! Ah'll take ye!' he shouted. The Gorbals children taunted him: 'Mad Billy, mad Billy, mad Billy, ye're aff yir heid!' On hearing this, the saliva came running out his mouth like he was a rabid dog. It was the cruellest of fates to be tortured not only by his injuries but also by children who could not understand his situation.

Another man who was taunted by the kids was the lavatory attendant at the public toilets in the centre of Gorbals Cross, near a large Victorian clock which was a local landmark. The attendant was a quiet, well-spoken man who kept himself to himself. He didn't drink or smoke. He always carried himself with a sense of dignity as if he was royalty. Because he was a silent type, many of the boys found it hard to work him out. After he had finished work at the toilets, he would walk along Gorbals Street in the usual dignified manner until one of the boys began to abuse him with the words, 'Shite man! Shite man! Shite man!' The local children also sang a street song, making sure he heard every word:

> My faither's a lavatory cleaner,
> He cleans oot the lavvies at night
> And when he gets hame in the mornin',
> His boots are aw covered in shite.

> Shine up yir buttons wi Brasso,
> It only costs tuppence a tin,
> Ye'll get it in Woolies fur nothin'
> As long as there's naebody in!

> Some say that he died of a fever,
> Some say he died of a fright
> But Ah know as sure as you know,
> He fell in a bucket of shite.

> Shine up yir buttons wi Brasso,
> It only costs tuppence a tin,
> Ye'll get it in Woolies fur nothin'
> As long as there's naebody in!

After six months in the job, he packed it in. The nightly psychological torture had been too much for him. He told me, 'Ah'm a quiet man who likes tae work for a livin'. People can be so vindictive here and life is hard enough withoot havin' tae put up wi' all that. Ah've got tae face the facts, Ah'm jist no strong enough for the Gorbals, there's nae dignity here.'

If he thought his life was undignified, perhaps he should have compared it with those of the jakes. These were extreme alcoholics who had been reduced to rags, in the gutter begging for money with their bunnets used as collection boxes. They were sad characters and, like Billy the boxer, didn't seem to have anyone around to help them out of their dire straits. The jakes gathered at the Clyde Suspension Bridge to drink hair lacquer or their favourite cocktail, 'jake': methylated spirits mixed with milk.

Terry the jake had once been a soldier with a family, a house and a future that looked promising. But, as he explained to me one day, 'A wee drink here and there led to a big drink here, there and everywhere. That was it, Ah lost the plot o' ma life.'

He was now reduced to begging with his bunnet for a drink that would eventually lead him into oblivion. At Christmas, like many other jakes, he wrote a seasonal message in chalk on the wall at his patch:

Christmas is coming, the goose is getting fat,
Please put a penny in the old man's hat.
If you haven't got a penny,
A ha'penny will do
And if you haven't got a ha'penny,
God bless you.

One day, Terry and his bunnet were no longer there. We were puzzled as to why he had gone missing. But too many jake cocktails had taken their toll. One of his fellow down-and-outs told us the story: 'Terry was awright wan minute, the next thing, he went aff his heid. He stood on the Suspension Bridge and shouted, "Ah'm signing aff noo. Ah've had enough," and then threw himself intae the Clyde. The polis fished him oot later, a few miles upstream, bobbing against the rocks.'

I went to see when the jakes threw a wake in a dark tunnel near the Suspension Bridge. I thought it was like a scene from hell: shabby, bearded men in dark overcoats sitting in the darkness drinking bottles of wine and jake. I was offered the jake cocktail but turned it down, laughing and covering my embarrassment by saying I didn't like milk. But I think I did realise then that they were modern-day lepers, left in seclusion to die in a Scotland that did not really care. Terry's best pal, Tommy, who had a hump on his back, poured the methylated spirits and milk down his throat and tears ran down his face as he sang 'We'll Meet Again'.

I felt extremely downhearted about it all and went back home to tell my father. He said we should all count our blessings and told me a saying that he had heard when he was growing up: 'I cried when I had no shoes, until I saw a man with no feet.'

'The Peanut Man' was another unfortunate character. He was an elderly, old-fashioned, sober kind of fellow and a genuine hard worker. He sat night in, night out in his little shop underneath the railway bridge in Cleland Street, lit by a single light bulb. It was no bigger than a cubicle. He sold roasted peanuts, tablet and candy apples. He looked like a product of the last century and he probably was. A grey-haired man, he always wore a dark, crumpled double-breasted suit with a grimy shirt and tie. Selling peanuts might have earned him peanuts, but he took pride in his job. How this man made a living trading in nuts from that dark hole always remained a mystery to me.

He became a victim of crueller children who went to his little front counter and kicked it up in the air. Or, as he sat there near his roast-peanut machine, they would throw stones at him, as if to persecute him for being a relic of a bygone age. The locals tended to support him, though, as he was a local landmark and part of Gorbals tradition.

The Peanut Man was a fellow of few words and I never got to know his name or where he came from or what his background was. One night when I passed by, the shop was closed down and in complete darkness. I never saw him again. In some sort of strange way, I missed his presence, because the Peanut Man had become a regular and expected part of my life. He represented the Gorbals of old. The Corporation later bricked up the peanut shop and people in the area thought it was an omen, the end of an era and a sign that the past was being bricked up forever.

Chapter 18

AN ALTERNATIVE EDUCATION

When I was 11, in 1967, it was time for me and my wayward pals to move on to secondary school. But the playground snake-belt fight came back to haunt me. At St Luke's Primary, the top ten brightest pupils in the final year went on to Holyrood Secondary, which was thought of as a very good school. The alternative was St Bonaventure's, a junior secondary where the not too bright went. Holyrood was considered to be a far superior establishment whose pupils went as far as university. By contrast, St Bonaventure's – 'Bonnies' – was not exactly known for turning out anyone with an academic pedigree. It did not even do O Levels or Highers. It had a monumental reputation for producing pupils who ended up on the wrong side of the law: hard cases, thieves, fly men, con men and the general element that continued to make up the population of the Gorbals underworld.

When I was at St Luke's, the headmistress called me into her room and said I was 'a borderline case'. She explained that although I had come ninth in the class, the snake-belt incident went against me. A school board meeting was held and it was decided that I was to go with all the other wild boys, and with some whom I considered to be dunces and morons, to St Bonaventure's. But they said with some good behaviour and academic results to match, I could eventually end up in Holyrood after my first year at Bonnies. It was a stitch-up and the headmistress knew it. She even gave me an Enid Blyton book as a consolation prize for being one of the best achievers in my year. I later threw the book in the bin. I was frightened but in a way I was glad, too, to be going to Bonnies, because all my rebellious pals were going there too. At least I would not be a stranger at my new school.

At that time, the headmaster at Bonnies, a little chubby man with a bald head, nicknamed 'Bud', was attempting to introduce stringent measures to improve the school's image. He even introduced a school uniform. This was a bit of a joke because many families could not afford one. And even if you got one, within a few weeks your school jacket was usually ripped off your back. Any uniform ended up in shreds because of all the fights that took place. At first, going to Bonnies was a terrifying experience. I was faced with the roughest youths you could imagine, far tougher than in most approved schools. There were fights most days and they were an extraordinary spectacle. Two boys would get into an argument in the playground and began fighting, then hundreds of spectators would gather in a circle around them. But they did not merely watch or heckle. Everyone began spitting profusely at the fighters. As the fight progressed, the opponents got covered in wave upon wave of spittle, drenching them completely. Surprisingly, the fighters did not take umbrage at this. To be spat upon carried some prestige; it was an experience, part of your journey through the supposed hard-man network. As the punches and kicks flew between one boy and another, there was a strange satisfaction in joining in on the spitting. At the end of the fight, the victor raised his arms up in the air; covered in blood, spittle and phlegm, he acted like a victorious Roman gladiator.

I had only been at the school for a few weeks when I had my first fight with another boy over an argument about football. We squared up to each other in the playground. As far as I was concerned, this lad would be easy to beat. When we started to slug it out, he managed to get a few digs in at me, but this was only because I was temporarily blinded by the enormous amount of spittle that was raining down. It was like fighting in a rainstorm. On the streets, I had been well trained for such combat. A quick Gorbals kiss and a kick between the legs resolved the matter in my favour. After emerging victorious and covered in most of the school's saliva, I vowed never to have a fight in the playground again. I had ended up in a disgusting mess. What would my mother say? That afternoon, I was on a Corporation bus heading home, covered in dried slobber, and a little notice stared down at me warning 'No spitting'. I thought to myself that I could have done with that proclamation around my neck only a few hours earlier.

What always intrigued me about Bonnies was the fact that it had the audacity to call itself an educational establishment. I never heard of anyone going to university from there, or any professor or genius claiming to have been at Bonnies. Some boys never went to the classes at all and had even been given special dispensation by the headmaster to shovel coal with the janitor in the boiler room. I suppose it was meant to be a kind of work experience but it seemed more like a form of slave labour to me. But if they helped the janny out by shovelling coal for a few hours a day, they were then free to play football in the playground for the rest of the time.

There was no great emphasis on learning anything academic. The class might be called mathematics, science or music, but the subject names were only there to persuade the education authority's school inspectors that all was well. The so-called lesson usually involved an extremely bored or inexperienced teacher handing the pupils a book and telling them to read it and keep quiet. Woodwork and metalwork were pushed as important subjects. It was presumed that because most of the pupils were working class, it was inevitable that they would go into the manual trades.

This lack of imagination also stretched to the education of the girls. They had separate classes and were based at the other end of the school. Mixed classes were considered 'unhealthy' and segregation of the sexes was the norm of the day. While we were being taught manual skills, like how to plane a block of wood or make a little metal garden shovel (which was ironic because none of us had a garden), the girls were being trained how to become instant housewives once they left school. Their classes included child care, cookery, dressmaking, home care, laundry and sick nursing. One girl even told us she had been to a class where they taught her how to make a bed properly.

Because of this unsatisfactory education, we had no real chance of going to Oxford or Cambridge or any other university for that matter, not unless we got a job there as a janitor or cleaner. When the teachers addressed the metalwork or joinery class, they usually said, 'When you leave this school, you have to get yourself a trade. Otherwise you'll be out of work and have no money. A trade is the best bet for your future.' Despite this advice, I honestly could not see myself grafting on some building site or in a factory. I was not alone: many of the boys considered a career in crime to be a far better alternative.

One day, the woodwork teacher began shouting at me, saying, 'You're no payin' attention. How the hell dae ye expect to get a job as a joiner if you're no payin' the least bit o' notice tae whit I'm tryin' tae teach ye?' I just said to him, 'Ah'm no the least bit interested in becomin' a joiner.' A look of disgust came over his face; he was outraged. I got six of the belt for that remark but although it was painful, I felt I was perfectly within my rights to say what I said. I thought he was trying to induct all the boys into some kind of working-class slavery. We were being trained to be small cogs in the machinery to make the capitalists money. Besides, I had watched older men who had been at the school and done the same things. Most of them ended up doing menial labour, or worse – in jail. The new headmaster at Bonnies tried to encourage some boys to take O levels but I never met anyone who sat for them. In the woodwork class, the only level I had was a spirit level. It was no good complaining about the lack of educational facilities; I did once, and the teacher said I was getting too cheeky and gave me three of the belt 'for my own good'.

The belt was frequently used to subdue the unhappy and restless pupils. There was a knack to taking the belt. Veterans like myself put our hands out and, just before the belt hit them, bent the fingers so that it hit only the fingertips and a part of the wrist. In those days, teachers had to buy their own straps. They sent away to John Dick, a saddler and tawse-maker in Lochgelly. They came in three weights, light, medium and heavy, and cost 7 s 6 d post paid. He supplied most of the belts used by teachers in Scotland. There was no discrimination; girls got the belt as well.

The so-called bad boys at Bonnies were often belted in front of a class full of girls. It was supposed to be a humiliation but it made them all the more determined to appear to be hard men and not cry. One particular teacher was nicknamed 'Scud' because he had a very thick belt that scudded sense into the pupils. He had been there for years and years and had belted most of the gangsters who now ruled the Gorbals. They all talked about him with a great deal of affection. Scud, a tall grey-haired man in his 50s who was always soberly dressed, in turn regaled us with tales about the characters who had passed through the school. It was clear he had a degree of respect for his former pupils. To him, the new generation of boys were just the same as all those he had dealt with before. He said to us one day, 'Ah've seen them come and go. All the fly men, hard men and chancers. Many of them still keep in contact. They now tell me that they needed tae be belted at the time, otherwise they would have went oot of control completely. They know Ah treated them hard but fair. The belt gave them that bit of discipline that they lacked in the house. It made them better people.'

Smoking at school was one of the offences you got belted for but this did not deter anyone. At playtime, up to 30 boys would be puffing away in the red-brick toilet block. There was continual talk among these smokers about the gangs, especially the ruling elite – the Cumbie. It was everybody's ambition to be a fully fledged member of the gang. Some even claimed that their brother or cousin was a leading member. The cigarettes and being supposedly Cumbie-connected enhanced their hard-man images but, ironically, I hardly ever saw any of them in a fight.

Perhaps this was because you had to be careful not to pick a fight with one of them as his brother or cousin might really be a member of a gang. I got it wrong big time. I picked on a boy who was boasting that his cousin was one of the leaders of the Young Young Cumbie, the gang's junior division. I taunted the boy, calling him a liar, and it ended up in a small skirmish. The next day, he was off school and I thought nothing of it. But the following day, his cousin and all of his pals, more than 20 boys, were outside. As I was coming out of the school gate, one of them jumped on my back and I hurtled to the ground. Then they all took great delight in giving me a good kicking. A couple of school pals helped carry me home. I spent a week off school with my injuries. My face and body were covered in boot marks. But I never revealed to the authorities, school or police, who was behind the attack. When I got back to school, I met up with the boy again and nothing was really said, but we did shake hands. He knew that I had got the message.

Although Bonnies was supposed to be for the ignoramus contingent, there were plenty of pupils at the school who had real talent. Take, for example, Jim. He was a brilliant drawer who copied the likes of Picasso, Dali and Matisse. When I watched him draw, I thought it was the work of a true genius. If you were lucky enough to be pally with him, he'd do a portrait of you. It was as good a likeness as any photograph could be. If Jim had been at a better school, he would have been hailed as a major talent and gone on to bigger and better things. Then there was Wee Francie whose life was one big song. He was another one everyone tipped to be a star in the future. He would stand up in the class when the teacher shouted, 'Hey, Francie, give us a song,' and belt out a variety of ballads. He had no musical education but every note he sang sounded faultless. He was a real one-man show. Instead of hanging around the streets, he sat at home and listened to Elvis Presley and Frank Sinatra, trying to capture the dynamism and charisma of their voices. Another fellow was a great writer and made up the most tremendous short stories. You could give him a theme and a short story or poem would be written right away. He read out his stories to the class and on one occasion the teacher was so impressed he told the boy to go to the headmaster's office, not to be disciplined but to read it out to him.

These talented Gorbals boys were only the tip of the iceberg. There were budding engineers, draughtsmen, designers, accountants, teachers, footballers – you name it. But most of them had one thing in common: after leaving school they all disappeared without trace. Some of them became drug addicts or alcoholics. Others spent years on the dole or became poorly paid labourers. Many ended up in prison and others in mental hospitals. Some were even kicked or stabbed to death. A whole generation of the most talented people in not only the Gorbals but Scotland gone to waste.

Bonnies was a conglomeration of the talented, the strong and the weak. The weaker boys went through hell trying to survive in such an environment. And when I look back at my Bonnies schooldays, I often think of the adage that what does not kill you makes you stronger. A vulnerable boy joined the school from another area and was taunted mercilessly, day in, day out, for months on end. He'd had a terrible upbringing: his father had died when he was young and he lived in a crumbling tenement with his brothers and sisters; his mother was almost never home. One day, he just cracked with one of the bigger boys and took him on. He got a bad beating. After a couple of weeks off school, he appeared again in the playground and was subjected to even more bullying and taunts. What happened next not only surprised me but shocked me. One of the lads called his mother 'an auld cow', meaning that she had low morals or was a prostitute. At school, this was considered the worst insult you could have. The boy's tortured face showed he had had enough. He pulled out a knife and plunged it into the bully's chest, shouting, 'Take that ya f****n' bastard!' Crying, and still grasping the bloodstained knife, he dashed out of the playground and ran along the road.

I didn't really think it was his fault, because he had been tormented into that situation. It was a real insight into how someone, particularly a weak person, could be forced into committing a murder. The next day's headline in the *Evening Citizen* was 'Schoolboy Stabbed in Playground'. It made Bonnies out to be the roughest school in Scotland. It probably was. While Holyrood Secondary was hitting the headlines for its pupils' success stories, we were making the news with a knifing.

Stabbings were not confined to that particular incident. One summer's day, a gang of teenagers aged 15 to 16 who were former pupils of the school turned up outside and were larking about. I sat with them joking in the sunshine and Chris, who was in the same class as me, joined us. There was a good bit of banter going on and a fight developed between Chris and a former pupil who had a reputation for being a tough guy. It ended up with Chris being stabbed in the arm. That incident didn't make any headlines, but it made me aware that at Bonnies violent behaviour and bloodshed were never far away.

Another lesson I learned at Bonnies was about drugs. My first experience of seeing how drugs affected people was when one of my classmates began smoking a lot of marijuana in the toilets. Alan, aged 14, was a short stocky lad who had scored what he told us was 'a good wee bit o' dope'. It gave off an incredible fragrance in the toilets, and he headed back to class well and truly stoned. Alan had dogged school one Friday to score drugs. If anyone played truant on a Friday afternoon, they would be given as a punishment three of the belt when they appeared back at school on the Monday morning. This was quite a good deterrent which cut down the Friday skiving rate. On the Monday morning, we waited with glee to see him getting the belt in front of us. We all knew he was terrified about it, even had a phobia. But after the class had enrolled, there was no sign of him.

A few minutes later, one of the boys rushed in, late as usual, and shouted, 'Sir! Alan's on fire!' We all rushed out into the corridor and there was Alan running about, a human fireball. We managed to get a fire extinguisher and some coats and doused him out. A badly burned Alan was rushed to hospital. Months later, he had recovered enough to tell us the story: 'Ah smoked a bit o' dope and came up wi an idea. Ah got some lighter fuel and put it oan my hands and set them on fire for a couple o' seconds until some blisters appeared. It meant Ah had an excuse tae avoid gettin' the belt, because Ah hate the belt. But the fire got oot o' control and Ah panicked. Ah tried tae put it oot by rubbin' my hands against ma jumper, but the jumper caught fire and the next minute Ah wis a ball o' flames.'

Most of the pupils at Bonnies were not really interested in drugs. We preferred buying a half-bottle of cheap red wine and then drinking it in the toilets during playtime. I found it quite fun being drunk during a lesson but the teacher often got suspicious because of the smell of wine in the air. This was easily counteracted, however, by buying strong spearmint chewing gum from the local corner shop.

Because the education provided was so mediocre and uninteresting, a great deal of contempt developed between pupils and teachers. Some weaker teachers found it completely impossible to control classes and just adopted a laissez-faire approach. One morning, I arrived late for registration and one of the strictest teachers in the school said to me, 'Why are you late, boy?' I felt nonchalant and extremely gallus replying in front of my classmates, 'Sir, Ah was oan a bus goin' through the Gorbals when it wis hijacked by a band of terrorists demandin' that they be taken to Cuba.' The whole class fell about laughing and the teacher was so dumbfounded he didn't even give me the belt. I had the feeling from then on that he had just given up on the whole charade.

A weaker teacher, who had a real plum in her mooth, had only been at Bonnies a few weeks before we could tell that she was on the verge of having a nervous breakdown. One day, she went out of the class for a couple of minutes, leaving her coat on her chair. When she came back, she shouted, 'Someone has stolen my purse from my coat. I'll leave the class for a minute and when I come back, I want to see the purse on my desk, and no more will be said about it.' I thought she was being a silly cow, as no one had been near her coat. She came back in a few minutes later. Realising that the purse had not reappeared, she made the same offer again before leaving the classroom. She returned to the same scenario. Then the headmaster was called in, offering the same deal, and then there was a mention of the police being summoned. The teachers went off and returned only to find the situation as it had been before. The next minute, the woman's face turned bright red and she told the headmaster that the purse had fallen down the lining in her coat. We innocent Gorbals boys had been accused of being thieves . . . terrible! Every lesson for weeks after that, she was too embarrassed to teach us. She let us play football instead.

My encounter with another teacher left me battered and bruised. As we were lining up to go into class, I said something impertinent to him and he punched me full force in the face, giving me a massive black eye; in fact, he almost broke my jaw. When I got back home with my swollen face, I told my mother, who went to the school the next day, saying she wanted to take the matter further. But the headmaster panicked and pleaded with her not to. He said the teacher was a good man at heart who was under a great deal of pressure and that he would lose his job if the authorities got involved. Later that day, the teacher made a grovelling apology to me, saying he had a wife and two children to keep. Losing his job would mean financial and professional ruin. How could he get a job again after being sacked for punching a child? To save the teacher's job, we decided not to pursue it. That did not stop him belting the hell out of me two months later after another run-in with him. My father commented wryly,

'That's one of the biggest lessons you'll ever learn: never give a sucker an even break.'

Another teacher, Miss McDonald, was a middle-aged spinster who tutored music in the most eccentric fashion. She taught us songs from obscure musicals but her favourites were the old Scottish standards like Robert Burns' 'Oh My Love's Like a Red, Red Rose'. At first, we all laughed and felt a bit embarrassed when she demanded we learn the words of such songs. It was not exactly the thing to do if you wanted to cultivate a streetwise image. But we did learn the words and ended up singing the songs because we thought it was a load of fun. Being in her class was like something straight out of *The Prime of Miss Jean Brodie*. She even drummed into us a musical version of the Ten Commandments, which I can remember, word for word, more than 30 years later:

> First, I must honour God;
> Second, honour his name;
> Third, keep holy this day truly,
> This will be my aim.
> Fourth, I must be obedient;
> Fifth, be kind and true;
> Sixth be pure in all I hear, say and do;
> Seventh, I must be honest;
> Eighth, be truthful in all things I say;
> Ninth, be pure in mind and heart
> And all I think and desire each day;
> Tenth, I must be satisfied,
> Come what welcome, come what may.
> These are God's Ten Commandments,
> These I must obey.

After school, back in the house, things were almost as mad and dramatic. My father had clearly been breaking some of the Ten Commandments. Every now and again, he went on one of his wild benders with his band of cronies, and it often resulted in some sort of trouble. One morning, as I was heading to school, the police came to the house and took him away for an alleged assault on an American sailor in the Broomielaw. He was bundled unceremoniously into a police car and driven off. I headed to school that day feeling really empty inside; I couldn't concentrate on my lessons and I thought I'd never see him again. When I came back from school, he was sitting quite cheerily in front of the coal fire. He said he had been freed because the police had a lack of witnesses and little evidence. Another drama unfolded when he was working away. He came into the house one night and told us that while he was making a phone call in the Central Station his suitcase had been stolen with all his working gear inside of it. It was lucky that he kept his money in his jacket, otherwise he would have lost everything. He blamed the theft on some of the unsavoury personalities who hung around the station. In the suitcase he had some of his chef's overalls and the treasured knives that he had collected over the years. They were worth a few bob but, because they were an essential part of my father's working life and the tools of his trade, they meant more to him than money.

A few weeks later the police arrived at our house to question my dad. He said the police had found the body of a man in England and he had in his possession my father's driving licence and other items belonging to him. When the police found out they had all come from the stolen suitcase, he was in the clear.

Dad was always in and out of jobs, probably because he still had a temperamental Glaswegian spirit. In one hotel, he was in a hot kitchen preparing the lunch for a 100-strong wedding party. A new young manager had just taken over and he walked into the kitchen and began tasting the dishes, saying things like, 'This needs a bit more salt . . . this needs a wee bit more seasoning . . . this sauce needs to be a bit thinner . . .' My father went over to the stoves, turned all the gas off, said, 'F*** you, you do it yourself,' and walked off.

The manager shouted, 'You can't do this, there are 100 wedding guests out there, you can't just walk out!'

'Who can't?' my father replied, throwing off his chef's gear and making for the nearest exit. Another job down the drain, but at least he had self-respect.

He landed a job in a hotel in Largs and had a similar dispute with another manager. The outcome was he stuck the nut on him, the police were called and he was promptly arrested and spent the weekend in the cells. Another day, another job.

For a while, he was extremely happy being the head chef in a hotel in Ayr. Me and my Gorbals pals would jump on the train from the Central Station to Ayr and stay overnight. Dad was in full flow and the owner of the hotel, a millionaire builder, had given him carte blanche to do whatever he wanted with the menu. Result? He invented Guinness soup and had the posh diners raving about it. He would laugh saying, 'Ah make a big pot o' soup and stick ten pints o' Guinness in it and the punters love it!' In many ways, watching my father going about his life was like watching a comic cabaret act; every moment seemed crazy yet exciting. After the hotel's restaurant had closed for the night, me and the boys would gather in his kitchen and he'd raid the alcohol larder, pulling out a bottle of Grand Marnier and pouring us all large glasses. It was fun. Once, to our amazement, he produced an air pistol and began firing it off all over the kitchen, exploding a few lights on the way. On another late night, we were all sitting around having a laugh when the singing duo the Alexander Brothers walked in. They had made a name for themselves in Scotland with a hit single called 'Nobody's Child'. It was well after midnight when the two kilted cabaret singers strolled into the kitchen. One of the brothers said, 'Chef, we've just finished a show at the Gaiety Theatre. Any chance of making us a couple of rounds of sandwiches?' Me and the boys began to giggle in expectation of my father's reaction (we all had glasses of unpronounceable spirits from the kitchen's cooking larder). Dad turned round and said, 'Would I come round tae your hoose at midnight and ask ye tae sing "Nobody's Child"? Get tae f***!'

Surprisingly, despite his numerous indiscretions, he lasted almost a year in that hotel. And then it was back to the Gorbals, looking for his next job. My mother would get mad at him saying, 'You know your trouble? Ye jist cannae haud down a joab, ye've got nae staying power.' My father merely shrugged his shoulders and replied, 'Aye, Ah'll tell ye whit, it's hard work lookin' fur hard work. Anyway, as the auld saying goes, he who follows the crowd is never followed.'

Chapter 19

CUMBIE YA BASS!

When people think of the Gorbals, they often have in mind the razor gangs of the 1930s and 1960s. The Gorbals was indeed home to a gang when I was growing up, a very powerful one. To some extent, this group of wild young men controlled what happened not only in the Gorbals but also in a large section of the city centre. The gang was called the Cumbie. It had originated years before in Cumberland Street. Older people recalled gangs before the Cumbie, like the 1950s Bee Hive, which had been formed outside a well-known shop of the same name, also in Cumberland Street.

The Cumbie were making headlines as early as 1960. The *Evening Citizen* reported that two of them were sent to jail after gang members, brandishing hatchets and bottles, 'cut a swathe of terror through the district'. A police spokesman said:

> There are about forty members of the Cumbie gang, and most of them live in the area. They hang about Cumberland Street and cause trouble. There are times when they try to extort money from betting shops with little luck. But they extort the occasional fiver or tenner from street bookies as protection.

The Cumbie legend began to really take off in the mid-'60s, when gang warfare became prevalent in the Gorbals. There were monumental fights on Friday nights, with razors and even swords used. When somebody shouted, 'Cumbie ya bass!' – the gang's war cry – it was guaranteed to send a shiver down the spine. It was usually after the pubs came out that the majority of violent dramas unfolded. There were head-buttings, kickings and slashings taking place. Some local people got a buzz out of the big fights because they knew most of the faces involved. Many of the gang members were trying to live up to some kind of hard-man reputation or another.

By 1967, there were three sections of the Cumbie: the Big Cumbie, otherwise known as the Caley Road Cumbie; the Young Young Cumbie, known as the YYC; and the Tiny Cumbie. The Big Cumbie was made up of a lot of mature people who were real gangsters. They carried not only knives but also pistols and shotguns. They were looked upon as being in the premier league of their division and were involved in bank robberies, professional violence and moneylending.

The YYC were all young guys in their late teens, the best fighters of their generation. They had gallus personalities and wore superb clothes which made them look like the most fashionable teenagers in Scotland. They wore Levi's Sta-Prest trousers and Doc Marten boots and had their shirts personally made by tailor Arthur Black in the town's St Enoch Square. The Tiny Cumbie were mostly boys in their early teens who looked up to the YYC. All of them were trying to make sure they'd made a name for themselves by the time they graduated to the YYC. In turn, the youths in the YYC looked up to the Big Cumbie, as that really was like going to gangster university. The Big Cumbie looked up to nobody. You don't have to when you are the top of your division.

For a young guy like myself, it was almost inevitable that I'd end up involved with the Cumbie. I went to the right schools and lived in the right street – my pedigree was considered ideal. By the time I was 14, most of my friends and I were members of the Tiny Cumbie.

If you wanted to spot the Cumbie, the best place most nights was outside of John the Indian's corner shop in Crown Street. There, dozens of the Tiny Cumbie gathered at night-time. The YYC, who were more gallus, fashionable and elusive, preferred propping up bars in and around St Enoch Square and elsewhere in the city centre. The Big Cumbie could be spotted in pubs all over the Gorbals wearing flash suits and doing the sorts of deals you wouldn't tell your grandmother about.

The YYC even had its own battle song. It went:

Standing on the corner on a Saturday night,
Up came some bams who wanted a fight
Ah pulled oot ma razor as quick as a flash
And shouted 'Young Cumbie, Young Cumbie,
Young Cumbie ya bass!'

There was also highly amusing graffiti. One gang member spray-painted a wall with a six-foot message proclaiming: 'EVEN THE DEAF HAVE HEARD OF THE CUMBIE'.

The Cumbie's biggest rivals were the ever-diminishing San Toi and the larger and stronger Calton Tongs. Most of the gang warfare developed over who actually controlled certain parts of the city. It was more or less divided into two: the Cumbie controlled the Gorbals and a large part of the city centre; the Tongs reigned over part of the city centre and from Glasgow Cross right up the Gallowgate. Like the Cumbie, the Calton Tongs were massive in size and in presence. The two rival gangs often fought in the city centre, especially at the halfway mark at Glasgow Cross or St Enoch Square. The Tongs had much the same structure as the Cumbie. They had come to prominence in the Gallowgate area of Glasgow after a group of boys saw a movie called *The Terror of the Tongs* featuring horror-movie star Christopher Lee. What appealed to the young gang members of the Gallowgate was that the movie was about a secret oriental society called the Tongs. The film was set in the gold-rush era in the USA and the oriental Tongs were very much like the Chinese Triads. After the movie was shown, young men began to riot in the streets of the Calton, shouting, 'Tongs ya bass!' As a result, Gorbals gang members shouted, 'Cumbie ya bass!' A Glasgow lawyer argued in court during one gang trial that 'ya bass' was in fact a polite French phrase. But his eloquence could not disguise the fact in Glaswegian terms it simply meant 'ya bastard'. In 1967, Gorbals shop owner Pat Lally, a Labour councillor, was strongly criticised for retailing blood-red ties boldly adorned with the words 'Ya Bass!' But he told the *Evening Citizen*, 'They are frivolous. Anyone who thinks otherwise is adopting pompous and moral attitudes.'

The Cumbie and the Tongs sometimes did battle in the Glasgow

Green when the carnival appeared there for the annual Fair Fortnight and hundreds of gang members went there armed to the teeth. But most of the time fights resulted from a stand-off situation. You knew your face was being clocked by rival gang members and any time you left the safety of the Gorbals to go into the city centre, a member of one of the rival gangs might recognise you and then you'd be in for it. Aged 15, I was standing in the Central Station one night waiting for a Cumbie acquaintance when four young, well-dressed members of the Tongs approached me. One of them said, 'Hey, pal! Do ye no remember me?' His face rang a bell and I could recall seeing him a few weeks before at the Glasgow Green. There had been a bit of a skirmish and he and the rest of his gang had run off. On that occasion the Tongs had been outmanoeuvred by the Cumbie.

He punched me in the side and then ran off. The punch didn't feel that painful but then somebody shouted, 'There's blood comin oot yir shirt!' It was then I realised I had been stabbed. A passer-by phoned for an ambulance and I was rushed off to the Southern General Hospital in Govan where I had been born 15 years before. The doctors said I was lucky to be alive as the knife had not punctured my heart or lung. It taught me a valuable lesson: always to be on the alert when I was in the centre of Glasgow. It really was a matter of life and death.

Gangs had long had a very powerful presence in Glasgow and had had a considerable impact on the city's culture. Catholic gangs in the Calton years before had become known as the Tim Malloys, rhyming slang for 'the boys'. Even to this day, Celtic supporters in Glasgow are known as Tims; indeed the phrase Tim Malloy is used by Rangers supporters to describe anything Catholic.

In the mid-'60s, one local politician, fearful of the Cumbie and other such gangs, said: 'The authorities have allowed a situation to develop where, if truth be known, people cannot walk along the streets.' The situation was such that the Scottish folk singer Hamish Imlach penned the song 'The Cumbie Boys'. The message of the song was that the Cumbie were a Catholic gang preoccupied with worshipping their new messiah, Jock Stein, while the rival Derry gang from Bridgeton were Protestant and hated the Pope. Imlach's advice was simple: it wasn't wise to wear a Celtic scarf in Bridgeton or a Rangers one in Cumberland Street unless you were a champion boxer or a very good sprinter.

Sociologists blamed the growing permissiveness of the 1960s for the upsurge of such gangs. They also looked at the historical factors, tracing them back to the 1840s and 1850s, when shiploads of Irish immigrants fleeing the potato famine landed in the West of Scotland. This resulted in the Catholic–Protestant divide in Glasgow, an aspect of life there which persists to some extent to this day.

All three branches of the Cumbie regularly joined together to form one giant gang when there was a Rangers and Celtic match on at Parkhead, Ibrox or Hampden. The most dangerous walk was between the Gorbals and Parkhead, during which they had to pass through the notorious Protestant 'blue-nosed' area of Bridgeton. But no Cumbie member was that afraid, as up to 200 men and boys would turn out – it was like being part of a private army. The great majority of the Cumbie were tooled up with hammers, razors, air pistols and antique swords. One Saturday afternoon before an Old Firm match, a rival gang, the Protestant Derry Boys, began a running battle with the Cumbie boys near Bridgeton Cross. There was head-butting and kicking and the waving of swords, but no one was seriously injured as hundreds of the Cumbie headed to Parkhead.

The newspapers started to publish stories saying that razor gangs similar to the ones in the 1930s were back. There were reports of youth clubs in the Gorbals and other areas of Glasgow suffering because their membership had fallen off, as most young men wanted the excitement of hanging about with the gangs. As early as May 1965, the crime figures for the Gorbals and the rest of Glasgow made alarming reading. They revealed that over 850 people had been arrested for carrying an offensive weapon. Also, more than 1,500 people had been apprehended for breach of the peace, and slightly fewer for disorderly behaviour.

In the Gorbals and the new outlying schemes like Easterhouse, local shopkeepers and residents expressed their concern over the emergence of such gangs. They urged the Lord Provost of Glasgow to do something about it. At one stage, there was even the preposterous spectacle of crooner Frankie Vaughan arriving in the city and asking the gang members in Easterhouse to give up their weapons, throwing them into bins for the cameras. In the media, the appeal might have looked quite successful, but to gangs like the Cumbie it was all a joke. There were plenty more weapons to go around.

By 1965, the national television and radio networks were already fed up of reporting the violence in Glasgow. They started to use phrases like 'a stabbing is no longer news in Glasgow'. A lot of the old men at the street corners in the Gorbals, ex-army types, watched the wild young Cumbie members parading down the streets and said, 'They should bring back the birch for these hooligans. In fact, flogging's too good for them.' This feeling filtered its way through to the May municipal elections, for which the self-styled Progressive Party printed posters urging the return of the birch. But some gang-member activists made their own in retaliation. The colourful posters were put up in shops and pubs and they boldly declared: 'Vote for your local gang – the Cumbie.'

If they weren't fighting with other gangs, the Cumbie were actually quite proficient at keeping law and order. For example, thieves who broke into working people's houses were given instant violent punishment. The Cumbie built up such a reputation that a lot of the older members had big IRA connections in Glasgow and in Ireland.

There were so many gang members getting injured at one point it became a bit of a standing joke. I heard that one of the Cumbie gang got into a taxi at Gorbals Cross and the driver said, 'That'll be thirty bob.' The guy said, 'How can it be thirty bob if ye don't know where Ah'm goin'?' The taxi driver replied, 'Ye're goin' tae the infirmary – ye've still got a hatchet stuck in yir heid.'

One night, we saw a guy at Gorbals Cross hobbling on a pair of sticks. We thought the invisible man had arrived in the area because his face and hands were completely covered in bandages. The ludicrous thing was he was wearing a pair of glasses over his facial bandages. One of the boys recognised him as a leading member of the Cumbie. He was, as Alex aptly said, 'a bigger state than New York'. We asked him, 'Hey, Jimmy, dae ye want a haun to where ye're gaun?' (Remember, at that time, every man in Glasgow was called 'Jimmy'.) He gave a grunt and pointed to a nearby pub. So we guided him to the door and his pals, as shocked as we were to see the mess he was in, helped him inside and sat him down. It turned out he had been drunk and mouthing off in the Gallowgate, the result being he had been attacked by the Tongs and left for dead in a puddle in a back lane. Somebody called for an ambulance and he was patched up at the hospital but when he heard the CID were coming up to interview him, he managed to crawl out of his hospital bed. Swathed in bandages, he was helped into a taxi which had taken him to Gorbals Cross.

Some of the young Gorbals guys went really over the top when it came to being armed. Frank, who lived with his bedridden mother in Nicholson Street, used to go out with a bayonet, razor, hammer and air pistol on him. He had been slashed the year before by a member of the Tongs in George Square and was taking no chances. The slashing made him mentally unstable, as before that he had been a good-looking fellow who was always a charmer with the girls. Because his movie-star looks had been ruined by a massive scar on his left cheek, he vowed vengeance. Frank told us one night, as he headed up the town, 'Ah'm well prepared for any eventuality and if they try tae get me again Ah'm gonnae stab, slash and shoot the bastards.' A few months later, he appeared in court for attempted murder after getting himself involved in a gang fight in Argyle Street. It cost him ten years of his life to protect himself. I thought it was a big price to pay for having a pint in the city centre.

Chapter 20

POLICING THE GORBALS

The Cumbie and other Glasgow gangs became so powerful that the police admitted they were fighting a losing battle. They began to get a bit nervous about stopping anyone in the street and carrying out a search, because if they did, they could end up being surrounded by a huge mob of young guys baying for their blood.

A chief inspector from Glasgow addressing a police conference in the mid-'60s said: 'I have mentioned the word "vandal". But this, like "juvenile delinquent", is a modern term. We in the police service prefer the original expression: "hooligan".' He added an ominous warning: 'Police officers are, thank goodness, human beings. They display a great deal of restraint. But how much longer they can hold themselves in check is another question. If the courts will not use the teeth provided for them, the police may decide to do the biting.'

The problem was that most of the time the Cumbie had far more teeth than the police officers. If two policemen were walking down Crown Street and there were fifty or so Cumbie boys, they often shouted, 'F*** the polis!' or the more humorous 'Funny polis!' Nine times out of ten, the policemen walked off in the other direction pretending that they hadn't heard them. Policemen were often subjected to drunken taunts and the V-sign, which might get you arrested or, if you had really offended them, beaten up. In desperation, they even started an advertising campaign declaring 'Help the Police', which was on posters all over the Gorbals. I couldn't stop laughing, though, when somebody wrote underneath the message on the posters: 'Kick f*** out of yourself!'

There was a general contempt for the police. Even when I was a young lad, people were always telling me things like, 'Never talk tae the polis, they're a right waste o' time. They don't deserve a nod in the desert.' When I was aged ten, one of my older relatives sat me down one night and gave me this advice: 'Don't tell the polis anything. It's a case of see no evil, hear no evil, speak no evil wi them. I mean, how can ye trust somebody who swears an oath tae the Queen and even says he'll arrest his mother and father if he sees them committin' a crime? The polis are a shower of bastards but unfortunately ye need them noo an' again.' He said the local police were 'as bent as the people they're arrestin'.' In a way, he was right. One night, we followed two police officers on patrol and they found a shop that had been broken into. We saw them taking their own haul and later stacking it in one of their lock-ups.

In 1969, one bent ex-policeman, Howard Wilson, shot two police officers after taking part in a bank raid. Wilson also tried to kill an inspector but his gun jammed. After leaving the police force, he had landed in a lot of debt, but an attempt to solve his immediate financial problems by opening up a greengrocer's in Allison Street did not pay off. He joined up with two other members of a gun club and decided to become a bank robber. In the 1960s, before the advent of modern technology, a lot of Gorbals guys believed this was the easiest route to fast money. Wilson and his accomplices came up with a half-baked idea to rob a bank in Linwood but were spotted soon afterwards as they returned to Wilson's flat above his shop in Allison Street. The shop was near Graigie Street Police Station (well known to all of the Gorbals' criminals and bevy merchants) and the local police had an inkling that Wilson was up to something and began to keep an eye on the comings and goings at the flat. Wilson must have believed he was invincible, choosing to base himself so near his old colleagues. The police eventually raided the flat but two of them were killed by the ex-cop. This was an utterly insane thing to do, as cop-killers were reviled but a cop killing another cop was unheard of. Wilson served 32 years in jail and, ironically, his greengrocer's shop thrived.

One afternoon, my friends and I were going down Hospital Street near the Clyde when we spotted two policemen throwing stones at a bulging sack they had just thrown into the river. We stood and watched as the sack burst open. It was full of cats trying to swim for their lives.

The policemen were hitting them with big stones so that they would drown quickly in the murky waters of the Clyde. I was a bit shocked, because in the house I had two cats and I considered this a cowardly and inhumane thing to do.

The policemen noticed that I had seen what they had done. They both came over to me. One of them grabbed me by the jumper and said, 'Ye never saw that and if ye tell any c**t, you and your daft pals will end up gettin' locked up. We'll make sure ye get sent tae approved school.' I replied, 'I never saw anythin', officer. Sure, I wouldnae tell anybody if Ah did.'

They let me go, slapping me on the back and telling me to be a good boy and behave myself. They believed that they had secured my silence. But as soon as I got back to Crown Street, I told everyone I could, even the biggest gossip in the area, Peter the paper man who, as I've mentioned, sold the *Daily Record* at the Bank Close in Crown Street. Soon the story was all over the Gorbals and I was glad the truth was out.

At times, the police moved in heavy-handed, with squads of them descending on the area, battering into the Cumbie boys and loading up their black Marias. But it was a futile task, as most of the Cumbie were out the next day after getting fined and were soon up to their old antics. Even when they were fined or remanded on bail, the Cumbie boys hardly had to bother because all the gang members usually chipped in and that cleared the debt straight away. If any member was arrested they usually refused to give a reply when charged at the police station. But if anyone did answer, the favourite line was: 'Ah'm sayin' nothin' until Ah see ma lawyer.' One gang member built up a unique reputation for getting found not proven at trial. When he was arrested and charged, he always uttered the words, 'I have been unjustly arrested,' and when he appeared in court he'd concoct a story around that line. I saw him in a gang fight near Gorbals Cross one night. When the police arrived on the scene, he ran off but was hotly pursued and arrested not long afterwards standing in a fish-and-chip shop queue. He protested his innocence, saying the police had mixed up him with somebody else as he was only out for a fish supper. His claim of having been unjustly arrested got him another not proven verdict at the High Court. If he had been found guilty, he would have got at least two years in jail.

By 1969, as a result of the escalation in gang warfare, the police had swamped the area. The gang fights were bad, but I also witnessed liberties being taken by the police. They arrested harmless wee drunken men for breach of the peace when they should have let them go with a warning. I saw the police beating people with their truncheons just because they had been cheeky and they didn't like their attitude. But, to be fair, the police must have had a terrible time. It was like being Wyatt Earp trying to control a Wild West town.

They were infuriated and a backlash by them was on the cards. The major criminals decided to lie low. They knew police resources were being poured into what the tabloids were describing as an uncontrollable situation. We realised it had become just too quiet. This hit me, Alex and Chris one day as we went on our usual stroll along Crown Street, through Gorbals Cross and then on to Eglinton Street. Every few blocks, we spotted a stranger standing in a close looking rather subdued but vigilant. Most of them did not look like Gorbals people, as they had beards and long hair. At first, we thought the Gorbals had been invaded by some sort of hippies, but on closer examination, these muscular, well-built men looked too aggressive for that.

One night, Alex was grabbed by a local hard case and battered in the mouth for being cheeky. A big hippy-looking fellow was standing in a close just a few yards away wearing a camouflage jacket. A few seconds after the incident occurred, with all the shouting and bawling going on, the big hippy ran over and broke it up. 'Police! You are under arrest!' he shouted. He was an undercover policeman and he pulled out a pair of handcuffs. He made one insignificant arrest for breach of the peace, but he had given the game away. From then on, strangers standing in closes, especially if they did not look or talk like the local inhabitants, were presumed to be policemen. One of the Cumbie said to us, 'You see these guys staunin' in street closes pretendin' they're wan o' the locals but they might as well wear uniforms, because we know that they're polis. And their problem is, they know that they're polis.'

The undercover police had had their cover blown and within a few weeks they disappeared from the closes. However, they had made plenty of arrests during their undercover crackdown and many people ended up appearing in the High Court, just over the Clyde in the Saltmarket. At the High Court, we watched our friends and relatives being tried for all sorts of offences.

The sentences handed out were particularly harsh, with many individuals getting five or even ten years for what we considered to be minor offences or trivial skirmishes. Long sentences were seen as a way of curbing the escalating violence; but some of the cases were blown way out of proportion. I watched such cases in silent disbelief. A policeman took to the dock and swore that he'd tell the truth, the whole truth and nothing but the truth. But he was in fact telling a pack of lies. I knew he was telling lies because I had actually seen the incident unfold. Some men who were considered locally to be just lowly bampots were made out to be top-notch hard cases and sentenced to long periods behind bars. A lot of the crimes resulted from the consumption of too much El D or Lanliq. One man was sentenced to life imprisonment for a murder that he could not even remember committing. As he was being led away from the dock, he shouted to the public gallery, 'Ah'm no guilty! It wisnae me that did it, it wis the wine!'

Older men who had served a fair bit of time said the young guys should not be afraid of going to jail, because if they went to Barlinnie, they would meet plenty of people they knew – it would be just like being back in the Gorbals. Oxford and Cambridge had their old boys' networks and the Gorbals had a similar system in operation, but the particular establishment was Barlinnie. It was a learning centre for all those who had made their careers out of crime.

Being in prison could change men and when they came out they often seemed different. Some were bolder and more aggressive, others timid and withdrawn; being in prison affected different people in different ways. I noticed that the men who had been in prison for a while smoked roll-ups that they called 'spiders' legs'. They were called that because while inside the men had learned the skill of rolling their cigarettes very thinly to save tobacco. Other guys came out still using jail patter and mannerisms. One common mannerism was for a guy to keep his hand behind his back when he talked to you. It was perhaps meant to signify that he might be concealing a weapon, and it was all supposed to be part of the hard-man image. More of a loser's image, we all thought. I remembered the story about the Gorbals con man who received a particularly harsh sentence for robbery, arrived at Barlinnie and said to one of the prison officers, 'Do me a favour, put me in beside the fly men.' To which the prison officer replied, 'There's nae fly men in here, pal, they're aw outside.'

The Gorbals children had a street song about Barlinnie:

A wannie go hame, a wannie go hame
Tae ma wee hoose in Barlinnie.
An when Ah get there, Ah'll pull up a chair
And wash ma face in a tinny

The purridge is great, ye don't need a plate
Ye only need a hammer and chisel.
A wannie go hame, a wannie go hame
Tae ma wee hoose in Barlinnie.

There was a great deal of respect between rival gangs for each other's leaders. When gang members ended up in prison together, there was hardly any fighting. Friendships developed as if they were all in a prisoner-of-war camp. It was a bit like the First World War story about the British and Germans giving up fighting to play football on Christmas Day. When all the rival factions were in Barlinnie, they were comparatively friendly. But when they got out of jail, it was war as usual.

Christmas Day was supposed to be a miserable time in jail but those who were on short sentences or awaiting bail saw Barlinnie as a kind of festive social club. There they could play ping-pong and snooker, have a turkey dinner and meet up with their old criminal acquaintances. But it was imperative that they got out for Hogmanay, as it was the big highlight of the year and a time to celebrate their freedom. We all chipped in to get the bail for one Cumbie gang member to be out for Christmas Day. He was none too pleased when he was let out, saying, 'Ah didnae mind being locked up over Christmas as long as ye would have got me oot fur Hogmanay.'

Giving evidence during a trial at Glasgow Sheriff Court in 1966, a

Barlinnie warder claimed that 40 per cent of the jail's inmates were either from the Gorbals or had relatives there. One local old-time criminal, Lenny, who had been in and out of Barlinnie all of his life since the 1920s, would get drunk on the wine and sing his favourite song, 'The Barlinnie Hotel'. Only old lags like him knew all the words. He would shout before starting the song, 'This is aboot the best hotel in Glesga. It costs nothin' tae stay there and it's never short o' customers!'

> In Glasgow's fair city, there's flashy hotels,
> They give board and lodgings to all the big swells,
> But the greatest of all now is still in full swing,
> Five beautiful mansions controlled by the king.
> There's bars on the windows and bells on the door,
> Dirty big guard beds attached to the floor.
> I know cause I've been there and, sure, I can tell,
> There's no place on earth like the Barlinnie Hotel.
>
> I was driven from the Sheriff and driven by bus,
> Drove through the streets with a terrible fuss,
> Drove through the streets like a gangster in state,
> And they never slowed up till they got to the gate.
> As we entered reception, they asked me my name,
> And asked my address and the reason I came.
> As I answered these questions, a screw rang the bell –
> It was time for my bath in the Barlinnie Hotel.
>
> After my bath, I was dressed like a doll,
> The screw said, 'Quick march, right into E Hall.'
> As I entered my flowery*, I looked round in vain –
> To think that three years here I had to remain.
> For breakfast next morning, I asked for an egg,
> The screw must have thought I was pulling his leg,
> For when he recovered, he let out a yell:
> 'Jailbirds don't lay eggs in the Barlinnie Hotel!'
>
> The day came for me when I had to depart,
> I was as sick as a dog, with joy in my heart,
> For the comfort was good and the service was swell,
> But I'll never return to the Barlinnie Hotel.
>
> ['flowery = flowery dell, prison slang for cell]

167

Chapter 21

WATCHING MOTORS

Shortly after we started at secondary school, my pals and I came up with an idea to make some cash. There were a lot of break-ins to cars outside of the Clelland Bar in Hospital Street, so we decided to start our own little business, offering to watch the cars for the punters. The Clelland was a pretty wild place in the mid-1960s. It was full of music and vibrancy and run by a big tall guy called John Rawley. It was a great time to set up our wee car protection business, because automobile ownership in Glasgow was really taking off. In those days, most people weren't bothered about drinking and driving, and the number of people who drove to the Clelland for a night out meant that at weekends the street outside was jam-packed with vehicles.

What happened was, a guy would drive up in his big flashy Ford Zephyr, for example, and we'd shout, 'Hey, mister! Can Ah watch your motor?' Inevitably, he replied, 'Aye,' and another customer, and a couple of shillings, was secured for the night. English lorry drivers also went there with enormous vehicles, mostly furniture lorries, that they slept in, parked just down the street. All in all, it was a nice little earner. Most of the drivers gave us a shilling or two bob at the end of the night for watching their car or lorry. This meant that on a Friday night we could end up with about fifteen shillings each – not bad money in those days. We saved quite a few cars from being stolen or broken into. As soon as we saw a potential thief or vandal, we waved inside the pub for the owner to come out. He would be only too glad that we had notified him.

One time, a businessman with a big Mercedes approached us and said, 'Some bastard has nicked my Mercedes badge fae the front o' my motor. See if ye can get me another one and I'll gie ye a few quid fur yir trouble.' The only other fellow we knew with a Mercedes was a wealthy Irish ex-boxer who owned a pub just down the road. We sneaked around in the darkness, unscrewed the Mercedes badge and made off back to the Clelland. The appreciative entrepreneur gave us a few pounds. A week later, the Irish publican approached us asking if we could get him a Mercedes badge as somebody had stolen his! Luckily enough, that night the businessman with the Mercedes was back at the Clelland. We stole the badge back off him and gave it back to the grateful Irish fellow, thus making another couple of pounds.

At the time, gangs of children all over Glasgow had turned to the watching motors game. It became so common, comedians in concert halls like the Glasgow Pavilion started telling jokes about it. The best one was about a man who pulls up in a flash car outside of a pub. A boy asks, 'Can I watch your motor, mister?' The man points to two snarling Alsatians in the car and says, 'I think those two can look after it just fine.' Later, the man comes back to his car to find four flat tyres and a sign on the windscreen saying, 'Hey, mister, can your dogs blow up tyres?'

Standing in the streets late at night also had its darker side, however. There were a lot of 'stoat-the-baws' cruising about in their cars, eyeing up and even trying to pick up youngsters. The phrase was our slang for paedophiles and derived from perverts watching children as they played, 'stoating the baw' being a game of bouncing a ball against a wall played by young girls. Wee Alex was propositioned by a stoat-the-baw in Thistle Street. The man said to him, 'Dae ye want tae go fur a wee walk, son? I'll make it worth yir while.' But Alex shouted over to his big brother and his gang, who rushed over, shouting at the man, 'Ya dirty f*****n' pervert! We're gonnae kill ye!' They gave him a kicking and the man's blood was everywhere, his smashed glasses in fragments on the street.

On another night, a man in his early 30s with a well-to-do Glasgow accent approached us in his large car and asked for directions to a street in Crosshill. Myself, Alex and a boy called Dougie pointed him in the direction he should go. The man then said, 'Do you want to jump in and show me the way?' He seemed quite insistent and I did not like the look of him, but the boys were game and quite keen to go for a wee hurl in his motor. They saw it as a bit of an adventure. However, I was reluctant, as I'd always been advised never to get into a car with a stranger. But on that night, Alex and Dougie did. I went back home, telling nobody. But a windae hinger had spotted the three of us talking to the man and the two other boys getting into the car. They did not return home and by four in the morning there was still no sign of them.

The top police from Glasgow CID banged on my door early in the morning and told my mother to get me out of bed. I was in my pyjamas, aged 11, with the cream of Scotland's detectives questioning me about what had happened. They continually suggested I was telling a pack of lies but I maintained I was not and gave them a full description of the creepy man and his car.

The boys were gone for several days and it became a national story. Their faces were all over the TV and newspapers. When the police did eventually find them, thankfully they were safe. Hundreds of reporters and photographers gathered in Thistle Street as the police took Alex and Dougie back home, with cameras flashing and the media spotlight firmly on them. Alex later told me the story. 'The guy drove us tae his posh hoose and tried to get oor troosers aff. But we told him tae get tae f*** so he locked us in a wee room for ages until his maw turned up and she called the polis.' It turned out that the man was mentally unstable, and he never faced any charges for child abduction. He was committed to Carstairs, a hospital for the criminally insane. It was a good job, because if some of the local men had got hold of him they would have faced a murder charge. After that, I noticed there were fewer stoat-the-baws cruising the area. The children had been warned by the police, their parents and their teachers to let an adult know if they saw anybody acting suspiciously.

Just around the corner from the Clelland was the Citizens' Theatre in Gorbals Street. When it was too cold and wet to watch the motors, we headed there and for a shilling we could sit in the gods, where it was nice and warm, and watch a play to pass the time. We saw productions like *Billy Liar*, *A Lily in Little India* and *The Resistible Rise of Arturo Ui*, a tale in which the main character was a gangster based on Hitler, with machine guns all over the place. The plays we saw for our shilling were sensational. It was stuff we never saw on television. Soon us boys, aged around 13, were able to talk in quite knowledgeable terms about European theatre. It gave us a form of education that very few street children ever experienced. When we went to school the day after seeing a performance, the English teacher was always taken aback. He thought we were a bunch of unenlightened barbarians, yet here we were having conversations about major plays, even quoting some of the dialogue. The dumbfounded teacher could not work out how we had acquired such intellectual knowledge and, to keep it as a running joke, we never told him.

The Citizens' had a studio theatre, which we thought of as its own drinking club, a few yards away, ironically, for a place frequented by the well-heeled, up a close. It was called the Close Theatre Club and it was always packed with luvvies and arty types. It staged smaller, often controversial productions until a fire wrecked the place in 1973. We often tried to sneak in there but, because of our young age and scruffy appearance, we were usually quickly ejected.

One night, we had just come out of the Citizens' Theatre when we turned the corner and bumped into singer and actor Adam Faith, who was starring in *Billy Liar*. He found it hard to understand what we were saying because of our accents and we were a bit overawed because Faith had been a big star in the early '60s. My mother had quite a few of his singles in the house. After we got his autograph, we headed round to the Clelland to watch motors. Being near the Citizens', we often saw the odd celebrity or two mingling on the busy street with the locals. We saw quite a few star faces, but they became a part of our everyday life and after a while we took seeing such distinguished people in our stride.

Because hundreds of posh cars were being parked near the Citizens' Theatre, we decided to ask the theatre-goers if we could watch their motors. But the people who went to the Citizens' were a completely different breed from the Glaswegian working classes who went to the Clelland Bar. When we asked these folk if we could watch their motors, they usually flatly turned us down. I speculated that the majority of them lived in the 'Spam Belt'. This meant they lived in posh houses and drove fancy cars but did not have much money and lived on Spam behind their net curtains. The Glasgow petty bourgeoisie were far more careful with their money than the working classes. Many of them were not willing to part with even a shilling to have their vehicles looked after.

One evening, a woman who looked like a glamorous actress type left the theatre and was going towards her Mini when Rab, one of the wilder boys we hung about with, spat on her fur coat for no apparent reason. She went crazy, climbed into the Mini, mounted the pavement and tried to run us down. She then sped off round the corner and we thought we had seen the last of her. But we were wrong. She was a real drama queen and drove back, again mounting the pavement, trying to kill us all. She was stopped by the police but when she showed them the greasy spittle on her expensive fur coat, they let her go. The police came over and threatened us all with arrest if it ever happened again. It was a drama in itself, and probably more entertaining than what had been performed on the Citizens' stage that evening.

Rab planned to wreak revenge by breaking into the cars near the Citizens'. At that time, it was very fashionable to have headrests, which could be bought for a couple of pounds each from any motoring shop, installed in your car. Rab got hold of a joiner's punch from his uncle and used the device to shatter car windows. We then took the headrests, which were in great demand on the black market. After we'd stolen five or six headrests, we took them to a fence, a woman who ran a café in Crown Street. She bought them for ten bob a time. The surprising thing was she was a policeman's wife and her husband turned a blind eye to the situation.

Her café had a small room at the back with a jukebox. Armed with the cash the owner had given us, we had a wee party, with ice-cream nougats and Irn Bru all round. We were joined by numerous young lassies, who jived around the jukebox with us. As the shocked middle classes came out of the Citizens' Theatre to find their car windows had been shattered into fragments, we were all dancing in the café to the Beatles song 'Revolution'. In our own naive minds, we thought that by breaking into cars and giving the middle classes a lesson about their apparent meanness we had contributed to the working-class revolution. They had refused to give us even a shilling to watch their motors but now they were faced with big bills for their broken car windows.

However, the headrest scam did not last long, because some of the people affected, including a politician, a teacher and even a judge, had friends in high places. The streets around the theatre were soon filled with CID and plain-clothes cops pretending to be passers-by. We knew anyway, because the woman in the café had been told by her constable husband, that: 'The polis will be swarming all over the place. If Ah wis them Ah wid keep well clear o' the streets aroon there until the heat is aff.' It was time to lie low for a while. But we felt pretty good about it because, although we were hardly big-time criminals, it was fun to feel we had a mole with inside police knowledge.

Chapter 22

SHADY DEALINGS

The Gorbals was a hive of commercial activity and apart from the many legitimate businesses making money, there were numerous illegal ones as well. At weekends, people left the pubs and headed for illicit drinking dens called shebeens. There were quite a few dotted around the Gorbals during the 1960s. Some of them were run by gangsters or other crooks who always had access to stolen alcohol. They were very profitable enterprises as Scots law at the time was very strict on public-house drinking hours. When the pubs shut, either at the prohibitive time of 9 p.m. or, from the late 60s, at 10 p.m., punters wanted a late-night bevy. If they didn't have friends with a carry-out and a party to go to, a shebeen was the logical choice. The shebeens were ordinary flats in tenements. It was usually possible to run up a slate which the drinker could square up at a later date when he had the money.

A guy called Tony ran a shebeen in Hospital Street. He was never short of customers who had an unquenchable thirst after the pubs had closed and he made a good living. But other drinking-den operators got jealous of his success. One night, as Tony's shebeen was in full swing, a petrol bomb flew through the window. His regulars fled down the stairs of the close, while Tony stayed behind in a vain attempt to stamp out the flames. Eventually, he was forced onto the window ledge as the fire engulfed the place. He leaped 30 feet to escape the raging inferno and broke both of his legs. That was the end of Tony's career as a shebeen operator, but others prospered due to his operation's demise and the illicit drinking dens carried on as if the Gorbals was Chicago during Prohibition.

The police occasionally raided such places but, on the other hand, some of them would call in for a late-night drink. I regularly spotted a particular policeman leaving a Thistle Street shebeen well the worse for drink. He would stagger to his police box in Rutherglen Road and sleep there for a few hours until the end of his shift, by which time he'd have sobered up. The shebeen owner was happy to give this policeman free refreshments, as it ensured he would never be arrested, fined and have all his stock confiscated.

People were often punished in this way for turning their homes into shebeens. For example, in 1956, a Gorbals man, John Kerr, appeared in the local paper after being fined for using his Nicholson Street flat as a shebeen. His wife commented, 'For a while everybody was happy. Some of the people around here said it was great. If they had a big heid on a Sunday, they were able to come here and get something to cure it.'

It was not only the shebeen owners who made loads of under-the-table cash but the moneylenders as well. The interest rates were variable depending on whom you did the deal with on the street but it was usually one shilling in the pound. This meant that if someone borrowed a pound one week, they had to pay back a guinea the next week. If they missed a payment, it usually went up to two bob in the pound. A moneylender from Crown Street explained the situation to us boys: 'It's like this: we offer very low rates of interest and we expect people to pay us back the next week. We're no like the banks, we don't seize your belongings or your hoose. But if ye try tae bump us for the money, you're gonnae end up wi a sore face or even worse. Remember, it's no the principle, it's the money.' The service provided was fast and efficient and if you did not abuse the system and the 'generosity' of the lenders, everything was hunky dory. In a way, they were forerunners of today's easy credit. People could become physically sick with worry because they owed too much.

Certain moneylenders resembled the old American gangsters in the movies, with flash suits, colourful ties, highly polished shoes, neat haircuts and gold rings. They were not miserly people; they often gave some of their brass to good causes and they helped out many a destitute widow. They were hard but fair and if you were not even-handed with them, they could be unmerciful. It was utter folly to try and pull off such a manoeuvre, because, as the saying goes, 'Ye cannae con a con man,' especially a Gorbals con man.

The alternative to borrowing money was to put your worldly possessions into the pawnbroker's and redeem them as soon as you could. Men often pawned their best suits to have a drink, then had them back out a few days later when they got paid. The pawnshop could be generous or mean. It all depended on the temperament of the man or woman behind the counter. Housewives regularly had to go there to help feed their families when times were hard, but some were disappointed at the pittance they were offered for their goods. One of the regulars to the pawn was old Betty, who relied on it to get by. But she did not always get the desired outcome. 'That pawnbroker is as tight as a duck's arse. He offered me jist two bob for ma wedding ring, miserable sod that he is,' she complained.

The Gorbals kids had various rhymes about the pawns, one of which went:

> Old Mother Riley at the pawnshop door,
> Bundle in her arms and baby on the floor.
> She asked for sixpence, she only got four,
> And she nearly pullt the hinges aff the pawnshop door.

Another favourite was:

> Ah'm no comin oot the noo,
> Ah'm no coming oot the noo,
> Ah'm very sorry, Lizzie Mackay, for disappointin' you.
> Ma mother's away wi ma claes tae the pawn
> Tae raise a bob or two
> An' Ah've jist a fur aroon ma neck,
> So Ah'm no comin oot the noo.

Particularly hard-up individuals might pawn their belongings and later sell the ticket for a nominal sum but this was looked upon as an act of pure desperation. To many, the three brass balls outside the shop signified bleak times but to others they represented a form of social security.

One money-making scheme that was going on at the time was seen by the Gorbals hard men as easy cash. Protection rackets were springing up all over Glasgow. They were influenced by what the American Mafia had been doing for years. Pubs were a particular target, especially if a new landlord or owner had taken over. The idea was simple. A couple of local hard men approached the pub landlord saying they could 'stop any trouble happenin' in yir pub for a couple o' pounds a week'. If the offer was refused, they turned to another plan of action. They sent in their associates to start arguments with each other over nothing, sending tables, chairs and glasses flying and ultimately wrecking the place.

The landlord himself could be threatened with a beating to be delivered by complete strangers. Nothing too serious, though, otherwise he could end up in hospital. That would be no good because he wouldn't be able to settle up. Most victims fell for the scam and ended up handing over the required fee to their so-called guardians. This scam should have made a lot of the Gorbals gangsters quite rich but they tended to spend it as fast as they got it. It was a case of easy come, easy go.

Some hoodlums used their money to set themselves up as pimps. They controlled gangs of prostitutes all over Glasgow. There was no shortage of women looking for that kind of work. They were glad that they had the heavy mob behind them just in case they came to harm from any of the punters.

Me and the boys often hung around in Hospital Street outside the Turf Bar, where we regularly saw this pretty young woman aged around 20. In fact, we all agreed that she was more than pretty – she was beautiful, with long blonde hair and a figure that most women would die for. We always said she should have been a model. But we knew she was on the game. We had the impertinence to ask her what she charged. Chewing gum and without batting an eyelid, she replied, 'A pound the lane, three pounds the motor and five pounds the hoose.' We all hee-hawed but we were aware that like all people in business she had a sliding rate of charges.

Betty's Bar in the Broomielaw was also a regular haunt of prostitutes. We often caught the ferry over to Betty's and offered to mind the cars and lorries outside. We knew most of the lorry drivers there from their visits to the Gorbals and they knew us, so it was quite easy to make a few shillings. The ladies of the night in Betty's Bar were also doing well, with English lorry drivers buying them drink and later using their services. The prostitutes were always friendly to us, asking things like, 'Ur yis awright, ma wee dolls?'

One of them was a lovely soul called Maria, who was maybe in her late 20s and drank too much. Sometimes she appeared to be in a state, either through drink or drugs, and her face was often black and blue. It was clear to us that someone was giving her regular beatings. One night, we saw her with a massive black eye. We asked her who was behind it but she just sighed and said, 'Ach, it disnae matter, son, he isnae worth talkin' aboot because he's a right bastard.'

Week by week, the bruises on Maria's face, arms and the rest of her body got worse. She came out of the pub one night with a big, rough-looking fellow in his late 30s. To us, he looked like a monster. Big bawface, beard, beer belly, wearing a woolly jumper. His rough appearance marked him out as an ex-seaman. In fact, he reminded us of Popeye's arch-enemy, Bluto, and every time we saw him we whispered to each other, 'Look, there's Bluto coming.' It was clear he did not like us boys; he snarled every time he spotted us. But, being streetwise, we were not that afraid of him. We thought that behind his menacing snarl he was really a bullying coward who would be easy to sort out.

Another prostitute told us that Bluto was a pimp who 'looked after' several of the girls and was also Maria's boyfriend. One Friday night, Bluto and Maria came out of the pub and they began having a violent argument. 'Why don't ye get tae f*** and leave me alane, ya big fat bastard?' she shouted. Bluto replied, with his eyes bulging out of his head, 'Shut up, ye're jist a dirty slag, Ah'll wring yir neck like a chicken.' He then smacked her right across the face, causing her to stumble and fall. We began shouting at him, 'Leave her alone, ya big liberty-taker.' He ran towards us, tried to kick me and missed but managed to skelp Alex across the head. We made off as Bluto helped Maria onto her feet then escorted her up the road. A few days later, we were standing once again outside of the pub when Bluto appeared, well the worse for wear. He began shouting and swearing at us. 'F*** aff, ya wee bastards, get back tae the Gorbals where ye belong, or Ah'll kill ye.' He picked up a brick and threw it, missing our heads by inches. We ran into the shadows of the Broomielaw, just glad we had escaped the wrath of Bluto.

The next week, we turned up again and asked one of Maria's friends how she was. She began crying. 'Oh, Maria's been done in – murdered!' she said. We asked who did it but she said, 'I know who it was and you know who it was and that's aw I'm saying.' Bluto was arrested on suspicion of her murder but had managed to arrange an alibi with two of his girls, who said he was with them on the night of the murder. He pointed the finger at a group of foreign sailors who had been seen in Betty's Bar with Maria. The police let him go and Bluto was free to be a pimp again.

The next time we saw him, he appeared friendly and gave us a wave, shouting, 'Awright, boys? How's it gaun?' Of course we weren't awright. We knew he had murdered Maria and we had to do something about it. We got back to the Gorbals and told one of the Cumbie's main leaders what had happened. He had a reputation as a real game guy and a bit of a psychopath. But he just shook his head and said, 'Whit can ye dae? It's nothing tae dae wi oor territory. Now, if it had happened in the Gorbals, it wid be a different matter.'

We thought it had been left at that. But a few nights later the same guy approached us with about 20 of his fellow gang members. 'Awright?' he said. 'Where's this Bluto character? Lead the way and we'll gie him a lesson he'll never forget.' We headed for the ferry and Betty's Bar, then waited across the road in the shadows. For almost an hour, there was no sign of Bluto. The Cumbie were about to give up and head back to the Gorbals when he suddenly appeared, walking down the street towards the pub. The gang ran over to him and we could see the fear on Bluto's face. One of the Cumbie shouted to him, 'Hey, big man, Ah heard yir good at doin' in wee women. How's aboot showing us how hard ye ur noo?' He turned to run off but one of the gang head-butted him to the ground, another slashed his face and the rest of them kicked into him. Bluto was left in a pool of blood, still bleeding profusely when an ambulance took him away unconscious.

We never saw him again at Betty's Bar but one of the girls said he had fled Glasgow and was too scared to return.

Back in the Gorbals, I noticed that certain members of the Big Cumbie and other local hoodlums often disappeared from the area for a while. At first, I thought they were doing spells in prison, but I later found out that they travelled all over the country doing 'wee jobs'. These varied from bank robberies to demanding money with menaces. Some of the hoods had connections in London, including the Kray twins, and they came back with stories about them, saying they were 'a couple o' nice Cockneys, real gentlemen'.

The 'Jocks' were respected by gangsters like the Krays because they were thought to be harder and more vicious than most of their English counterparts. For example, when Cumbie hard man and debt enforcer Jimmy Boyle went on the run from the police in Glasgow in 1967 he headed straight for London to work for the Krays, who were said to have provided a safe house for him. But he was tracked down by a posse of Glasgow CID, who arrested Boyle in a pub which was frequented by leading members of the underworld. The Krays even came to Glasgow for the weekend in the mid-1960s. They talked about setting up some sort of operation but they found that the locals were too aggressive, especially after they'd heard the brothers' English accents. Another London gangster, 'Mad' Frankie Fraser, who had Glasgow underworld connections, stayed in the Gorbals for a short time and, in a recent autobiography, said he had nothing but praise for the hospitality of the people.

London criminals were sometimes brought up to Scotland by the top Glasgow gangster Arthur Thompson, who was said to rule the city. There were numerous anecdotes about Thompson and the moral was always that you never, ever tangled with him. He and fellow Gorbals gangster Paddy Meehan had attempted to blow open a vault in Inverness-shire in 1955 but both were apprehended. Meehan, because of his long list of previous convictions, got six years, while Thompson got three.

Thompson had begun to make a name for himself in the early '60s after being credited with shooting and beating up rival gangsters. Some of the Gorbals guys hung around Thompson's two clubs, the Raven

Club and the Hanover Club, in the city centre. In the mid-'60s, a fellow on a visit to one of the clubs was slashed repeatedly about the face. Thompson was charged and appeared at the High Court for assault but had the case dropped because of lack of witnesses. The police later managed to close down his clubs on the pretext of enforcing fire and safety rules.

Thompson also had an illegal street bookie's business, for which the Gorbals enforcers worked. He ran it quite openly and his enforcers told us that Thompson actually wanted a conviction for running an illegal bookmaking business, as he could then explain away any large amounts of money he might be found with in the future. One thing was for sure, whenever we saw Thompson in the Gorbals, we knew something was happening, and a lot of folk got a buzz from actually seeing him in person.

It was often said that crime did not pay. But from what we could see, it paid very well. There were a lot of prosperous-looking people doing it. Take, for example, a fellow called Jack who moved into a tenement in Thistle Street. He was always well turned out and usually had big wads of money on him. He was a thin, jumpy-looking guy who chain-smoked incessantly. He was very amiable with us Gorbals boys and often asked, 'How ye doin', pals?' as he passed us by on the street corner. One time he even stopped to have a bit of banter with us. After a few minutes, he said, 'Ah've got tae rush boys, because while the monkeys laugh in the trees, the lion moves on!' and with that he gave out the most tremendous roar of laughter. It was like we had a long-running secret joke with him. We knew that he was up to something and he knew he was up to something but nobody mentioned a thing – we just laughed.

His wife, though, seemed more serious about life. She was a ferocious-looking Irish woman who always had a worried expression on her face. Albert commented wryly that she always looked as though somebody had 'just done a shite in her handbag'. But they had two children, a boy and girl, who looked happy enough.

One dark foggy night, we were walking up Thistle Street and Jack, puffing and panting, asked us to help him up his tenement stairs with two large bags he had fetched from a van. As we were going up the stairs, one of the bags burst open and, much to our surprise, hundreds of pound notes fell out. He shouted, 'Hurry up, hurry up for f***'s sake.' We helped him stuff the notes back into the bag. It didn't take a genius to work out that he had obtained this money by illegal means. He told us to keep quiet about the incident, especially if the police were snooping about, and that he would make it worth our while. The next day, someone showed us a headline in the *Evening Citizen* which said that a big bank robbery had taken place in Glasgow. There was more than £20,000 involved. The problem was that Jack was a good-time Charlie. Once he got a drink in him, he was full of Dutch courage and began flashing wads of money around in the pubs. He was spotted buying even complete strangers rounds of drinks and no one could recall his ever having a proper job. A few days later, the police raided his house and he was taken away for questioning. He said nothing and they found nothing. But his young daughter told us in the street, in broad daylight, that he had hidden some of the money, packed in a plastic carrier bag, in the lavatory cistern and the police had failed to find it. 'The polis thought they had ma da bang tae rights,' she said, 'but they wurnae fly enough tae search the ootside lavvy.'

We later heard that his wife had left him because, as their next-door neighbour told us, 'His missus couldnae stand the pressure of being married tae a bank robber. It's no like being married to a joiner or a plumber. The money is far better but that disnae make up for aw the heartache and stress ye suffer. And because o' his trade he's always in and oot o' jail. That's nae life fur a woman wi two weans. Robbin' banks is a single man's job.' She was right: Jack later got ten years for another bank robbery, which ruled him out of being a typical Gorbals family man.

Chapter 23

ROMANCING AND DANCING

When spring came to the Gorbals, romance was usually in the air. The street boys who gathered on the corner were joined by wild young lassies who were full of outrageous patter. They dressed in bright-coloured clothes, smoked endless cigarettes and could be as troublesome as the boys. They were known locally as 'wee herries'. Some of these young darlings even carried steel combs to use in clashes with other girls, which were usually over a pimply youth. In 1968, a TV crew came to the Gorbals, capturing the wee herries in action for a documentary that was later shown on local TV. The programme showed them singing in the street:

Yummie, yummie, yummie,
We're the Soo Side Cumbie,
And we feel like chibbing the Tongs.

They formed their own gang, called the She Cumbie. This wasn't uncommon: other gangs in Glasgow also had their female counterparts, such as the She Tongs in the Calton. I was there one night when about 50 of the She Cumbie clashed with the same number of She Tongs in the city centre. The scene was enough to take your breath away. Punches and kicks abounded, hair was being pulled out in large clumps, there was scratching and biting and the steel combs were used. After the fight had ended, all the battered and bruised herries headed off home to lick their wounds. A taxi driver who had been in his cab near where the fight had taken place said to us, 'You won't find many ladies among that rabble, but they're women aw the same. Wan o' these days they'll be mothers, aunties, grannies, and naebody will believe what they got up to when they were younger.' He was right enough. Years later, these ferocious lassies blossomed into women. They now lead respectable lives in Glasgow and all over the world, surrounded by their children and grandchildren, with their steel combs buried deep in the shadows of time.

As we got older, we began to experiment with alcohol and sex. It was the usual stuff – half-bottles of cheap Four Crown wine and cans of Tennents lager. Wee parties developed in the dark recesses of the back courts; teenage girls turned up to have a swig or two and then winchin' sessions went on for hours. A few girls ended up pregnant and some families started that way.

Certain girls were so promiscuous they were known as 'line-up merchants'. One particular girl, nicknamed 'Charlotte the Harlot', often took on up to a dozen boys, one after the other, in a dark lane. One night, the police arrived in the lane and found ten boys there. They noticed that the alley was littered with condoms and all the guilty boys burst out laughing when a policeman said to a red-faced Charlotte: 'Where do you work, lassie? Are you a salesgirl at the Durex factory?'

Boys were given the contraceptives by their big brothers or older friends. The quickest and easiest way to get them was to go to the barber's. After giving you a haircut, he always said, 'Do you require anything for the weekend, sir?' Why barbers sold condoms always baffled us but I suppose it was an old tradition. When one of the first contraceptive machines arrived in the toilets of a Gorbals pub, one wag wrote on it, 'Buy me and stop one.'

If we didn't 'meet a bird' on the street corner, going dancing was the logical option. Traditionally, many people went to the Barrowland Dance Hall in the Gallowgate. It had a high success rate when it came to looking for a partner. My aunties used to tell me about their dancing days. If they met a young man at the Barrowland, my granny would say to them, 'Does he take a drink?' If they said no, my granny said, 'Never trust a man who disnae have a drink.'

The other two dance halls frequented by Gorbals people were the Plaza Ballroom at Eglinton Toll and the rougher Portland Dance Hall in South Portland Street. The Plaza had an old-time atmosphere, with a large dance floor, a revolving glitter ball and nice, small, intimate tables where men and women could chat to the potential loves of their lives over a drink or two. By contrast, the Portland was a far more rough and ready establishment. Dozens of gang members hung about there, joined by burly Irish labourers and every fly man imaginable. Plucking up my courage, I went there at the tender age of 14 and was taken aback at how uncivilised it was. One man was urinating near the stage as couples danced.

The Portland was a haunt of major members of the IRA, some of whom were highly active gun, drugs and explosives smugglers. The bands were mostly southern Irish, with acts like Big Tom and the Mainliners, who played a variety of Irish tunes, and the atmosphere was like being in an Irish town. At the end of the night, everyone had to stand to attention when 'A Soldier's Song', the Irish national anthem, was played. If you did not do so, the rules were simple: you got beaten up by the bouncers. The secret was to show respect for the implanted Irish culture. On another visit to the Portland, when I was 15, I was obviously not paying much respect, as I was punched so hard in the face by a drunken Irish fellow that I was lucky not to lose all of my teeth. The women at the Portland were less sophisticated and harder drinking than those who went to the Plaza, but many relationships and marriages developed between people who met at the Portland, until it was reduced, like most of the Gorbals, to a pile of rubble in the 1970s.

When romances began to flourish, there always had to be a designated meeting place. The most popular rendezvous spot was outside of Boots the Chemist at the corner of Argyle Street and Union Street. Every Friday night, dozens of young men and women met up there and then set off for a night out. But every Friday night, there would be some girl standing there greeting her eyes out because her fancy man had not turned up. This was called 'giving someone a dizzy' and the spot was known as Dizzy Corner. 'Ah cannae believe it,' a pretty, tear-stained girl said to us one night. 'Ah knew he wis a wee chancer when Ah agreed tae go oot on a date wi him and he hisnae even bothered tae turn up – he's given me a right dizzy. Wait till ma big brother finds oot. He'll kick his heid in, the wee bastard. Naebody gie's me a dizzy and gets away wi it.'

If any guy was heading to the dance halls at the weekend, it was important not only to dress up in a natty set of clothes but also to have a haircut that made you stand out from the crowd. There was a phase in the 1960s when the Gorbals became awash with Hollywood star haircuts. This meant either a 'shed cut' – a smart style cut with a razor – or a 'Tony Curtis' – short at the sides and long on top with a quiff. There was a joke doing the rounds at the time: a young fellow said to an old Gorbals barber, 'Gie's a Tony Curtis.' After a few minutes, he looked in the mirror and was shocked to see the barber had shaved all his hair off. He shouted, 'Whit ye doing? I asked ye for a Tony Curtis!' The old barber replied, 'Wis that no that guy in *The King and I?'*

Haircuts varied from the Tony Curtis to the crew cut, which had a military look about it, to the bowl cut, which was worst. It identified you as being so poor you had your hair cut in the house using a pudding bowl. An Italian barber called Felix in Rutherglen Road did the best haircuts. He used to complain to us, though, that local gang members kept breaking into his shop to steal his shaving razors to use in fights. When he was cutting my hair, Felix kept me and his other customers highly amused by telling his favourite anecdotes. I think he must have nicked most of his jokes from the fun section in the *Sunday Post*.

'A fella came in last week and Ah said, "Dae ye want yir hair cut roon the back?" And he says, "No, Felix, it's raining outside."'

'Another fella came in and said, "How much for a shave?" I says, "Two bob." And he says, "How much for a haircut?" I says, "Five bob." He thinks for a minute and says, "Ah'll tell ye whit, shave ma heid."'

Throughout the 1960s and into the early '70s, pop music became an increasingly important part of the Glasgow dance and party scene. Most teenagers bought a single every week from their wages. One local guy, Andy, who was in his early 30s and lived alone, had collected hundreds of records and as teenagers we used to go round to listen to them. The ironic thing was that Andy had only an old battered gramophone to play them on. But his collection was fantastic and he had records by just about everyone on the pop scene. Andy wasn't a drinker or a smoker and didn't go to the dance halls, so all his money went on pop music. When he wasn't working as a labourer, he played his singles day and night. We often popped round to his house to have a listen to all the latest sounds, including the Beatles, the Rolling Stones and the Kinks. He said to me, 'Ah don't need a bird – pop music is ma girlfriend. It disnae nag ye and it's always there when ye need it. Ma music gives me mair pleasure than any woman could.'

I thought this to be an astute observation. That is, until we noticed that a local girl had taken a fancy to the record fanatic and his collection. Beryl, in her late 20s, lived down the stairs from Andy and he had invited her up to listen to his vinyl pile. She wasn't particularly pretty but we could see she was determined to get her man. Alex commented that she had 'a face like a well-skelpt arse and a body like the Michelin Man'. And Albert quipped, 'He calls her his melancholy baby. That's because she's got a heid like a melon and a face like a collie.' Beryl confided to us, 'Ah've been lookin' fur a fella fur ages that likes the same kind o' music as me. Ah love the Rolling Stones and the Beatles, and Andy loves them as well. We've got so much in common. Ah think Ah've found ma perfect man.' The words had us alarmed.

Some time later, Andy moved out of his tenement flat and we heard he had got married to Beryl after discovering she was 'in the puddin' club'. We didn't see him for about ten months because he had moved to another part of Glasgow. One day, Albert and I bumped into him in Cleland Street. We asked him how his record collection was going and he looked crestfallen. 'Ah hid to sell it,' Andy said. 'We've jist hid a baby, so Beryl said ma collection hid tae go, cause it's no cheap bringing up weans nowadays.' The money from his records had gone on a brand-new pram and cot plus other baby things. For a moment, he looked like a defeated man but then added, with a wry smile and a shrug of his shoulders, 'But Ah suppose that's rock and roll fur ye.' From then on, we all agreed that when we got older we would never get married.

Chapter 24

WINE TIME

A gang of us had begun to hang around the street corners and back courts at night drinking cheap wine like El D and Lanliq. But even as teenagers, we realised that the inexpensive vino gave us all a personality transplant. I had seen the movie *Dr Jekyll and Mr Hyde* and thought that wine like El D had the same effect as Dr Jekyll's potion: one minute we were well dressed and well behaved, then after a few swigs we turned into dishevelled, ranting monsters. All sorts of stupid things happened in the Gorbals because of cheap wine. It was absolute madness, with boys and men smashing shop windows, sticking the heid on people and committing other acts of vandalism. They'd wake up in a police cell with a dreadful hangover, covered in blood, their clothes ripped, with no recollection of the night before. They usually ended up being given a fine, with the judge warning them about their future behaviour.

I remember hearing a song that summed up the situation. The parody goes to the tune of Bacharach and David's 'I'll Never Fall in Love Again':

> What dae ye get if you drink the wine?
> A ten-pound fine and a year's probation,
> Get yir heid kicked in, in the polis station,
> Ah-ah-ah'll never drink the wine again.

There was no shortage of do-gooders in the Gorbals making valiant attempts at trying to stop folk drinking. People attended Salvation Army meetings in church halls, where they were encouraged to give up alcohol by signing the Pledge. They sang a song at the start of every meeting:

I promise here by the Grace divine
To drink no spirits, ale or wine,
Nor will I buy or sell to give
Strong drink to others while I live.
For my own good this Pledge I take
But also for my neighbour's sake,
And this my strong resolve shall be:
No drink, no drink, no drink for me!

Stories were told, sometimes accompanied by a slide show depicting a working man or woman on the way home with their wages being persuaded by their friends to go into a pub for a drink. They'd eventually become addicted to alcohol, spending all their money, their family suffering in poverty. The person would be shown resorting to robbing his kid's piggy bank for a drink. The end of the story always had a happy ending. The poor individual would be saved outside of the pub by an officer of the Salvation Army, who led them away from the evils of drink and back to their family. The meetings closed with the members singing:

You'll never find me in a public house,
In a public house, in a public house,
Oh, no! No! No!

By the age of 14, myself and most people of my age were getting heavily into the cheap wine, which led almost all of us into trouble. The wine gave me a big mouth and I got too arrogant and aggressive for my own good. I purposely started an argument with two giant Highlanders at Gorbals Cross, asking them: 'Who you looking at? Dae ye want tae fight?' One of them promptly knocked me out. I woke up about 15 minutes later wondering what the hell had happened. One of the boys said, 'Man, that wine really had ye going. Ye decided tae take on those big navvies and they were twice yir size. Ye were oot cauld. The problem is ye had too much wine and ye didnae know whether it wis New Year or New York. Ye're lucky that big fella jist knocked ye oot. Looking at the size o' him, he could have killed ye. He wis a giant in wellies.'

One night, I was walking along Argyle Street when a few Gorbals teenagers joined me with a carry-oot of wine. We were all drunk when we spotted a bright-red cowboy shirt on a dummy in a shop window. There was no way any of us could ever have afforded this expensive fashion item. One of the lads said, 'F*** it, let's get that smashin'-looking shirt.' He picked up a brick and threw it through the window. The pane shattered into a thousand pieces and I grabbed the dummy and staggered off towards the Clydeside. But two policemen appeared from nowhere and began a chase. I threw the dummy into the Clyde, thinking in my wine-fuelled state, 'What a waste of a lovely shirt.' I was soon apprehended. As I was sitting in the back of a panda car with handcuffs on, the police were convinced at first that I was on drugs. But one of them began laughing, saying: 'Whit kind o' eejit steals a dummy fae a shop windae? Ye cannae be right in the heid.'

He was completely right. As the effects of the wine began to wear off, I realised that it must have turned me temporarily insane for me to have even considered stealing a shop-window dummy. So the message was, if you wanted to stay as Dr Jekyll and avoid Mr Hyde, keep away from the strong wine. It was the safest bet.

While us boys favoured the cheap wine, many of the older men were hooked on whisky. As Mrs Carey, a neighbour who had lived in the Gorbals all her life, told me, 'Whisky drinkers never hiv any money and it's their families that suffer because o' it. Mind you, ye cannae get away fae the smell o' whisky in the Gorbals, it's in the air aw the time.' She was right: apart from the numerous pubs emanating booze fumes, there was a big whisky distillery in Ballater Street and it used to emit an incredible smell of barley, hops and malt that often overwhelmed the Gorbals. The smell got right up your nose and you could even taste it. At times I could even smell it on my clothes.

Even in the 1960s, the whisky producers did not use lorries for transportation but big Clydesdale horses and carts with an army of men driving them. Things had hardly changed since Victorian times as the whisky carts trundled through the streets. For the fellows who worked on the horse and carts, transporting whisky through the Gorbals was no easy task. Just loading a cask onto a rack took half a dozen men and the excise man always seemed to be monitoring them. Because of the stringent Scots laws and regulations, whisky was treated as liquid gold and no employee was allowed to enter a warehouse on his own.

One day, a whisky horse and cart stopped at Gorbals Cross and we asked the driver what the secret of making good whisky was. He said, 'It's aw tae dae wi the wood. If you keep the whisky in a good oak cask over the years, it jist gets better and better. The flavour comes through the wood.' He even had a rhyme that summed up the art of whisky making. It went to the old tune of 'Dem Bones':

> The brewer's connected to the mash man,
> The mash man's connected to the still man,
> The still man's connected to the warehouseman . . .

In my fertile young imagination, the warehouseman was connected to every publican in the Gorbals, the publican was connected to the guy staggering through Gorbals Cross, the guy was connected to the police, the police were connected to the courts and the courts were connected to the jails. People got into terrible states through the whisky. Whole lives disappeared into the bottom of a bottle.

The whisky firms employed brainy men from our area to work in their various offices in St Enoch Square just over the bridge. Teacher's and Grant's had offices there and employed several local men as whisky stock clerks. We spotted one guy we knew going into his office and had a peek inside. We couldn't stop laughing because there was no sign of modern technology, not a computer, calculator or even a typewriter.

Everything was being written in old ledgers lying on sloping desks. Some of the clerks reminded us of Bob Cratchit from Dickens' *A Christmas Carol* and we expected to see Tiny Tim or Scrooge walk in at any minute. We watched the top man, who was called the master blender, go about his work. His samples room was just a table and cupboard with whisky bottles in it. He tried out the various blends then washed his glasses in the sink. The ironic thing was, these men were in middle-class jobs but they probably earned less than the average plumber or joiner. They tended to be careful with their wages and we rarely saw them throwing their money away in pubs. But perhaps this was because they worked in the trade – they had a bonus in the form of buckshee whisky.

A lot of Gorbals men would have taken up a lower-paid job if they had free whisky thrown in, as that's where most of their wages went. One story that was doing the rounds at the time was of the Gorbals fellow who was spending all his money on whisky and it had effectively ruined his life. He picked up his hawf of whisky, looked at it and said, 'See you, Ah cannae see my wife because o' you. Because o' you, Ah cannae see my children. Because o' you, ma business went bankrupt. And because o' you, Ah ended up lying in the gutter, clad in rags. But I'll tell ye wan thing . . . I'll gie ye wan mair chance.'

A drinker who had done himself rather well at the bar but didn't show the effects of his libations was described in complimentary terms as 'hauding it well' or 'having a good heid on him'. Gorbals pubs were, overall, cheerful places and the bars echoed to such pleasant enquiries as 'Whit ur ye fur?', 'Ur ye fur annurra?' or 'D'ye want a big yin?'

If a fellow was going downhill rapidly by letting the alcohol take over he'd soon hear one of his relatives or pals saying, 'You'd better start screwing the nut or you'll end up in the model.' The threat was usually uttered by a wife, sister or mother worried about the man's heavy drinking. After that, the fellow would generally take heed and put himself back on the rails again. But the unfortunates who did not take this astute advice and carried on drinking might indeed end up splitting from their families and living in the model – a model lodging house. They were for men who had hit hard times, most of whom had a drink problem. The majority of these places were run by the Salvation Army and the biggest model in the city was the Great Eastern Hotel, which housed hundreds of such characters.

Through whisky, cheap wine and other strong drink, they had landed themselves in that position. The sad thing was the majority of them showed no sign of intending ever to give up. My father said he was passing the Great Eastern Hotel one day when an old alcoholic stopped him and said, 'Could you loan me a shilling, pal?' He duly gave him the shilling and asked the man what he was going to do with it. He replied 'Whit Ah dae wi ma money is ma business.' My grandfather, a former merchant seaman, said that in his youth he had slept in a model and at that time they did not provide beds. All the men hung over a rope to sleep and in the morning the rope was cut, the sleeping men falling to the floor.

The alternative to the model was 'lobby dossing'. Homeless men who kipped in the closes of tenements were called 'lobby dossers' and there was even a cartoon character in the *Evening Times* called Lobey Dosser. Thousands of men dossed in lobbies all over Glasgow and magistrates came down hard on them, imposing heavy fines and even jail sentences. Sometimes when I was heading to school in the morning, a lobby dosser would be lying in the close reeking of the night before's alcohol.

When people woke up with the inevitable hangover, they sorted themselves out by buying a bottle of Irn Bru. It was concocted in Glasgow in 1901 by Robert Fulton Barr. Every Saturday morning, hundreds of Gorbals men with thumping heads and parched throats cried out for their favourite antidote. I'd hear them shout to their long-suffering wives, 'My heid's nippin' and ma throat's as dry as the bottom o' a parrot's cage. Get me a bottle of Irn Bru . . . quick!'

One public house in Gorbals Street had a sign on its wall that said it all about excessive consumption: 'Alcohol is a great servant but a terrible master.' And as Mrs Carey observed, 'Everybody in the Gorbals likes a drink but naebody likes a Gorbals drunk.'

Chapter 25

DAFT BOYS

On Christmas Eve 1969, a group of us teenagers, all aged 14 and upwards and all members of the Tiny Cumbie by now, went to midnight mass at St Luke's in Ballater Street. It was unusual as, for a change, all the guys dropped their aggressive behaviour and got into a spiritual mood. Inside the chapel, we all sang hymns together. Outside the chapel, around one in the morning, thirty or so of the gang were having a bit of banter when a baby-faced police officer turned up and began asking all sorts of questions. He had made several mistakes. The first one was that he was on his own. Second, he had a Glasgow accent that did not sound half as hard as a Gorbals one. Third, he was vastly outnumbered. Also, he looked as young as us. But putting on a policeman's helmet had given him a sense of power and superiority.

As the gang stood around not far from the chapel, he asked, 'What are you boys doing hanging about here? What are you up to?' He asked the questions in a cheeky, sneering tone. Any fool could tell straight away that he had made a grave miscalculation. Here he was, addressing the courageous Tiny Cumbie like a gang of bampots. I was taken aback, though, when one of the boys, Tam, head-butted the policeman right in the face. Then some of the gang began booting him along the street as if he was a human football. I was a bit alarmed, thinking they might kick him to death, and urged them to stop in case we all ended up facing a murder charge. Afterwards, we made off, leaving him unconscious on the pavement. I thought what had happened had been completely out of order; but the young policeman had brought trouble on himself. He must have been half cracked, not to mention totally inexperienced, to have tackled one of Glasgow's wildest gangs alone, especially when we weren't causing any bother. As I walked back to the house in Crown Street, a drunk man bellowed over to me, 'Merry Christmas, son!' But I could only think that it hadn't exactly been a Merry Christmas for the policeman.

In the early weeks of the new year, there were several violent confrontations between the two big gangs, the Cumbie and the Tongs. During one such skirmish, a squad of police cars arrived at the scene and twenty or so gang members were arrested for breach of the peace and mobbing and rioting. Some of the Tiny Cumbie members were given three months' detention in a youth offenders institution, and even borstal. The Sheriff was particularly harsh, saying it was time to give the street gangs a well-deserved lesson. One boy used his parents as witnesses. They testified in the Sheriff Court that their wee 14-year-old boy had been walking his dog when the police arrested him for no apparent reason. I had seen the boy concerned, with not a dog in sight – he didn't even own a dog – wildly waving a hatchet. But he looked like a cherub before the bench with his hair neatly combed and nice shirt and tie. His mother and father were brilliant liars and the evidence they gave was more fantasy than fact. The cherub was cleared of all charges. I met him a few days later in Ballater Street and he was joking, 'Ma ma and da should hiv been actors. Aw the boys are saying they should hiv got an Oscar fur that performance.' I could only agree with him.

Around the same time, I was heading to a café in Gorbals Street for a pint of milk for my mother when I bumped into another adolescent gang member called Joe, who had been drinking all day. I was earnestly trying to keep a low profile and stay in the house more to keep away from all the madness. I had promised my mother that I would keep out of trouble. We exchanged friendly greetings and Joe followed me into the café, where he began shouting and bawling. He was like a madman, threatening the Italian owner and then kicking over the front counter. I had just paid for my milk when the police arrived and arrested Joe and, much to my surprise, me for breach of the peace. I was locked up in the local police station. My mother came to bail me out the next day. She had to pawn her one family heirloom, an old gold bracelet, to raise the funds.

I explained to her that the incident had nothing to do with me and that I had been wrongly apprehended by the police. Perhaps they had arrested me because, although I had only innocently popped out for a pint of milk, I looked every inch the Cumbie gang member with my Levi trousers and Arthur Black tailor-made shirt. And as the saying went, 'If it walks like a duck, looks like a duck and talks like a duck, it's a duck.'

I met Joe a few days later and he was very apologetic. He said that he would take the rap for everything and explain to the court that I was completely innocent. The case was due to appear at Govan Magistrates' Court and I was quite apprehensive about it. Despite assurances from Joe, I was convinced that if things went wrong, I would end up in approved school for buying a pint of milk.

With extreme trepidation, I turned up at the court alone, and Joe was nowhere to be seen. I thought he must have done a bunk and still put in my plea of not guilty. The Procurator Fiscal was a stout man with an upper-class accent. What he said took me aback. 'I am afraid I am unable to go ahead with this matter today as I have been informed by the police that the main instigator was kicked to death last week in a gang fight.' The magistrate dismissed all charges against me. I walked through the centre of Govan on a cold winter's morning thankful to be free and even more grateful just to be alive.

But the madness continued. A few days later, I bumped into a couple of the Tiny Cumbie late at night. They had just come out of the Portland Dance Hall. As we walked past a pub that had closed a few hours before, one of them kicked the door and it flew open. They ran in to get some booze and I thought it was best for me to disappear. A loud alarm went off and the place began to swarm with police. I ran into a dark close and hid for a few minutes. But a policeman spotted me, dragged me out and placed my hands on the roof of the police car in an arrest position. This made my hands very dirty. Even though I had not broken into the pub, they alleged I was part of the burglary gang. My dirty hands were used as 'evidence' that I had helped force the door open. Despite giving a strong defence, I was found guilty and the Sheriff remanded me for social-work reports, which for many boys ultimately meant three years' approved school.

Even though I was only 14, I was transported to Barlinnie Prison and put in overnight with some of Scotland's most hardened criminals. First of all, they put me into a little cubicle called a dog box. It was covered in graffiti written by the jailbirds over the years. Then I was put into a gloomy prison cell. Barlinnie was jam-packed with all sorts of characters, young and old, and it was like a town in itself. I saw many familiar Gorbals faces in there. In the darkness of the night prisoners defecated into pieces of paper and chucked them out of their cell windows. The next day, other prisoners, whose self-respect must have really suffered, had to pick up the soiled papers from the jail yard and put them into bins.

Myself and the two boys who'd kicked the pub door in were later transferred by prison van to Longriggend Remand Centre, near Airdrie in Lanarkshire, where we spent two weeks having social-work reports done on us. The place was extremely menacing. Numerous fights went on between the inmates. We were so young that we were ordered to go to the schoolroom inside the remand centre every day. The boys were rough with violent tendencies (which of course was what had landed them there in the first place). One morning in the classroom, I got talking to a boy who came from another tough part of Glasgow. For no apparent reason, he turned round and stabbed me in the chest with a newly sharpened pencil. I believe that if he had had a knife on him, he would have used that.

Social workers came to interview me, asking about my family background and whether I was likely to get into trouble in the future, which was ironic, as I should never have been arrested in the first place. After two weeks, I appeared back in court where, after reading the reports, the Sheriff decided I should be given one last chance.

Over the years, I saw many of my former classmates from Longriggend appear in the newspapers accused of all sorts of scandalous crimes. The boy who attacked me with the pencil became a murderer a couple of years later. He stabbed his victim to death with a large commando knife. I was glad he had used only a pencil on me. The tip of that pencil is still embedded in my chest and I often use it, when I look in the mirror, as a reminder of the 'good old days'.

Back in the Gorbals, the trouble continued and the crime wave went from bad to worse. For many of my peers, thieving had become a full-time job. When shoplifters had to get rid of stolen goods, there was no scarcity of fences to buy them. One, called Bert, always offered good money for the right gear and he had no problem finding customers to sell it to. Baby equipment and household goods were top of the agenda.

We went to his house and it was an Aladdin's cave stuffed with stolen goods. He used his front room as his stockroom. Most of the stuff had already been sold in advance. When the boys took their gear to Bert, he gave them roughly half what it was worth. He was quite an entrepreneur and also ran an illegal book at the local pub. Often, he didn't have to pay out cash to the winners because his punters would take goods in payment instead – all stolen, of course.

One night when he was at the pub, someone broke into his house. But, as a precaution, Bert had bought a ferocious Alsatian and as soon as the burglar entered the house the massive canine sunk its teeth into his leg and would not let go. When Bert got back, he was greeted by the sight of the guy lying on the floor moaning as the Alsatian kept a tight grip with its teeth. Of course, Bert was in a bit of a dilemma as to what he should do. He could not exactly call the police. After making a couple of death threats, Bert got a couple of his cohorts to help carry the injured guy to a close across the road and an ambulance was called. One of Bert's associates got into the ambulance with the burglar to make sure he kept his mouth shut when he got to the hospital. When he arrived, the injured fellow told the medical staff that he had been attacked by a rabid dog. They had no problem believing him – after all, it had happened in the Gorbals. Bert laughed when he told the story, and said, 'Serves him right, that's a lesson fur him: ye should never ever try tae steal fae a thief, especially wan wi a big dug!'

Bert had every reason to laugh, because he was doing a roaring trade; many boys my age had taken to breaking into shops. Their method was simple: kick the door hard until it flew open, walk away, then come back shortly afterwards if no one appeared on the scene. Inside the shop, they helped themselves to as much as they could carry. Those who had the most nerve in such situations made the most money. Because hundreds of buildings were being demolished in the Gorbals, another way to break in was to find a shop under an empty tenement. They then sawed through the first-floor flat's floorboards so that they could drop down into the shop.

This plan was thought to be almost perfect, as a policeman could be outside not knowing the gang was inside helping themselves. Another method was to get a hacksaw and saw through the bars of the back window of a shop. Boys who could not get hold of a saw would find a large block of wood and simply bend the bars before squeezing through the window. Cash always came before goods. Some shops left dummy cash in their tills, with the real cash hidden somewhere else. So when a group got into a shop, the hunt would be on for the real cash and if it was found in a great enough quantity, most of the goods were usually left where they were.

There were 15-year-old boys with more money in their pockets than grown men who had been labouring all week on building sites. The Gorbals boys, loaded with cash from their robbing exploits, became cocky and boasted quite openly: 'I've got mair money on me than my da and he did 12 hours' overtime last week. Working is a real mug's game.' But they later found to their cost that screwing shops could also be a mug's game. Many of them landed in approved school or borstal, because you were either a lucky thief or an unlucky one. I knew of boys who had burgled just one shop being caught and sent down; by contrast, other boys broke into hundreds of shops without ever getting caught. If you were one of the lucky thieves, everyone wanted to be your friend and join you in your adventures.

Child crime in the Gorbals was nothing new. One elderly guy showed us a newspaper cutting from 1949 reporting on an eight-year-old boy who was admonished on a charge of breaking into a Gorbals bank with intent to steal. He was described as 'not so much an infant Raffles as a child imbued with the spirit of *Alice in Wonderland*'.

The police had a suspicion that many of the burglars getting away with their crimes were from Bonnies and they had a quiet word with the headmaster, who then had a word with the teachers. As a result, one of our teachers stood up in class and said quite bluntly, 'I know some of you are oot there screwin' shops. If you don't stop it, ye're gonnae end up in jail. So you'd better pack it in noo, otherwise ye face a long time behind bars.' But the words fell mostly on deaf ears. The pupils had to snigger, because one particular young teacher was always buying stolen goods to give to his girlfriend.

Those who ended up in approved school hardly put others off. They described a relatively comfortable existence which sometimes involved scrubbing floors and polishing brasses. They reckoned doing time in approved school was a doddle compared to living with their families in a cramped, damp single end in the Gorbals. They often arrived on home leave with their pals from approved school. These young strangers to the area were overawed not only by the legend of the place but by the intensity of life in the Gorbals. They felt they had entered an alien world where normal rules and regulations went out of the window.

Of all the boys, the one who seemed most likely to become a gangster was Big Brian. He had a good pedigree: all his uncles had been involved in gangs. Brian himself had proved to be a strong fighter and an accomplished thief. At the age of seventeen, he had managed to avoid approved school but had served three months' detention then nine months in borstal. When he came out, he looked super-fit and he obliterated opponents in fights. These fights lasted only a few minutes because of his superb physical condition. Brian was a bit more than wild, though; there was talk of a 'mad streak' in the family. His mother was a bit odd and his uncle, a wild guy in the Big Cumbie, was known for his crazy exploits. His uncle was sitting in a pub one time when a group of rival gang members walked in. He simply pulled his bunnet over his face and said, 'If Ah cannae see them, they cannae see me.' And, for some reason or other, his policy of sticking his head in the sand worked.

Brian had the same mad streak. If we were walking through the city centre on a Friday night, out of the blue he'd mug someone on the spot, demanding, 'Gie's yir pay packet or I'll chib ye.' He also had a reputation for shoplifting from department stores. If caught, he put up a fight, and he always got away. He'd boast, 'Ah'm game as f***!' We had to agree with this observation. His waywardness could be hilarious at times but he got to a stage where I thought he had gone completely round the bend. He called a few of the gang together for a meeting and said he had planned a bank robbery. 'There's a right few bob in this,' he told us. 'Wan o' the bank staff carries the money fae wan branch tae another and it'll be a cinch to take it aff the tube. It'll be easy money, nae bother at aw. Like taking candy fae a baby.'

I turned down the offer to join them. The local papers at that time showed that the judges were handing out stiff sentences of up to 15 years to bank robbers. But the mugging went ahead and worked according to plan. Brian and a few of the boys ended up with thousands of pounds. The problem was, he had such an uncontrollable personality he could not keep his mouth shut and he was eventually arrested for the robbery. He was given a lengthy term of imprisonment. The next thing I heard, he had got into a fight with some prison guards in Peterhead. He got such a bad doing that he went even more crazy and was later sent to Carstairs mental hospital. Years later, I got the shock of my life when I met him in the street. His mind had gone and he could barely string a sentence together. He had been given electric shock treatment and that, coupled with the beating and the numerous drugs he was forced to take in the hospital, had reduced him to a shell of a man. He was like a sad, demented character from a Gorbals version of *One Flew Over the Cuckoo's Nest*.

Of course, with all this criminal activity going on, the Gorbals was swarming with probation officers; but they were usually fighting a losing battle. If someone had got into trouble, the probation officer not only saw them at his office in the city centre but also visited them at home. During home visits, boys behaved themselves. The house would be spick and span as the youth's mother made the probation officer a nice wee cup of tea and gave him a slice of fruit cake.

The conversation usually involved the mother saying, 'Ah've told him Ah don't know how many times tae keep away fae they bad boys.' Every mother said the same things to the probation officers and they just nodded their heads in approval in the same way every time. My friends in the Gorbals were not really bad boys who had led me astray.

We were all as bad as one another and only had ourselves to blame.

The probation officers came to visit adult criminals as well, mostly members of the Big Cumbie, who had been convicted of all sorts of crimes too numerous to mention. If a probation officer arrived unannounced at the front door, these guys would usually pretend not to be in and go to ground until they were summoned by letter to appear at the office. But seeing a probation officer was far better than being in prison, so they had to play the game, at least up to a point. As one experienced criminal explained to us, 'Yir probation officer can be yir best pal or yir worst enemy. The secret is tae only see him when ye hiv tae. Then make sure ye're aw pally wi them, otherwise ye're back in jail before yir feet can touch the ground.'

The Gorbals was also teeming with academics having a go at being social workers. To me, they all looked the same: beatnik or hippy types with beards, glasses, woolly jumpers and posh Scottish or English accents. These bohemians saw the Gorbals as a sort of missionary paradise where they could write intellectual dissertations on the social deprivation of the place. They also questioned people like myself for hours about what it was really like to be a part of the Gorbals community. I thought that it was like being a laboratory rat undergoing experiments. Indeed, one of them asked me to keep a diary outlining my diet and sleeping patterns, and even gave me daily graphs to fill in. After a while, myself and the rest of the boys ended up treating them with the utmost contempt calling them 'a bunch of posh diddies' who would have been better off helping the starving of Africa or India rather than pestering the people of the Gorbals. We speculated that when they went home to their middle-class environments they blew their own trumpets about the dreadfulness of working in the neighbourhood. But I suppose it helped many of them get a PhD. We thought that, for them, the main thing was that being involved with the world-famous Gorbals, with its connotations of poverty, alcoholism and violence, looked good on their CVs.

Chapter 26

TRUANT ADVENTURES

In my final year at school, in 1970, most pupils were so disillusioned that they simply did not bother to turn up. A boy in my class came from a heavy-drinking Gorbals family from Thistle Street, who habitually had wine parties in the morning and through the afternoon. So instead of going to school, we just hung about his house watching all these inebriated characters having a sing-song and telling the most astounding and bizarre stories. A woman called Cath, who was known to be a real wine-mopper, used to be there. She had a teenage son who was considered to be an outstanding footballer. 'The manager o' Glasgow Rangers came tae ma wee single end and said he wanted tae sign up ma boy. He asked where his faither was,' she said, 'but Ah hid tae tell him, "The boy hisnae got a faither. He's no a bastard, he's a love child."' The boy was indeed a tremendous footballer but he never made it to the big time because he inherited his mother's love for wine.

Another big drinker, a grizzled old navvy nicknamed 'Puncher' because in his younger days he was known to punch people out, had us in fits with his storytelling technique. 'When Ah wis a boy ma faither gave me a shillin' fur a pie. Ah went tae the bakery shop and saw a tray o' pies in the windae. Ah pulled ma shillin' oot but then it fell doon the gratin' intae the bakery where they made the pies. Ah wis greetin' and when Ah got back hame ma faither felt sorry fur me and gave me another shillin'. Ah went intae the shop and bought another pie. And when Ah bit intae it, ye know whit wis inside?'

'The shilling?' I said.

'No. Mince,' Puncher replied.

It was an education in itself.

By late afternoon, most of the party would have fallen asleep and several of the boys would rifle through everyone's pockets for money. I remember Alex laughing and clutching a handful of small change, saying, 'Crime never goes tae sleep!'

When we were dogging school, we often went up the town and skipped into the pictures. One afternoon, I arrived in the city centre alone after taking the morning off school, went up a side lane and prised open the door to the cinema. The movie was on and I could see only one person sitting there in the darkness. From the back he looked familiar and it turned out it was Alex, who had also dogged school and skipped in. Why the management didn't notice baffled me even then. We were the only two people in the picture house and neither of us had paid a penny to get in.

Playing truant meant that we couldn't risk staying in the Gorbals where we might be spotted during school hours, which led to many escapades all over Glasgow. We might venture over to the West End and visit the Kelvingrove Art Gallery, which would keep us occupied for most of the day. To avoid being detected as a dogger, it was best to mix in with parties of school children who were visiting the place on educational trips. The teachers at our school had never taken us there, so when we went we thought we were creating our own enlightening excursion as we gazed at the masterpieces, including Salvador Dali's *Christ of St John on the Cross*.

We also went to the Museum of Transport, which at that time was in the old tram depot on Albert Drive, not far from the Gorbals, and spent hours looking intently at the old motor cars, steam trains, trams, buses and models of all sorts of ships. Entrance to the place was, as for most museums in Glasgow, free, which was ideal if we decided to sidestep school on a rainy day. A favourite game in the motor-car section there was 'bagsing' something. That is, if you saw a car you fancied, you said, 'I bags that,' meaning that was your particular wish for the future. I bagsed an old Rolls-Royce, while Alex bagsed a Bentley. We laughed when one boy, who hadn't much imagination, bagsed the Hillman Imp. Years later, I saw him driving an old Hillman Imp through Gorbals Cross. He tooted his horn at me and shouted through the window, 'Hey,

Colin, ma bags came true, yours didnae!' As I wasn't behind the wheel of a Rolls-Royce, I couldn't exactly disagree with his observation.

A good way to get around town was to hop aboard the Glasgow Underground, or the subway, as we used to call it. Back in the 1960s, the old system had barely changed since it first opened in 1896. It was the only underground system in the UK outside of London and trundled around in a circle, taking in 15 stations. Climbing aboard the old Victorian trains meant that we had the whole of Glasgow before us. Our journey started at Bridge Street in the Gorbals then took us on a merry-go-round ride through West Street, Shields Road, Kinning Park, Cessnock, Ibrox, Govan, Partick, Kelvinhall, Hillhead, Kelvinbridge, St George's Cross, Cowcaddens, Buchanan Street, St Enoch Square and then back to Bridge Street.

It was a fantastic way to see how the residents of Glasgow, rich and poor, ticked as they went about their daily lives. When we alighted at stops such as Kelvinbridge or Hillhead, we noticed straight away how prosperous it was compared to the Gorbals. The locals talked in upper-class accents but they did not make us green with envy. I thought they were like aliens compared to the people of the Gorbals. Although we lived in comparative deprivation, we never ever thought we had bad lives. We had lives full of adventure and intrigue. The middle classes may have had their wealth but we didn't consider their lives to be half as exhilarating as ours. The division between classes was deep. Residents of well-to-do suburbs like Hillhead and Kelvinside simply did not mix socially with the common working classes. My grandfather told me a story that illustrates this point. Two ladies from Kelvinside were on a city-centre shopping expedition when they accidentally dropped into a less than hygienic café for afternoon tea. On the table in front of them were the remains of someone's lunch. 'Look at that, Marion,' said the one lady to the other, 'someone must have lost their appetite.' A wee Gorbals waitress heard this and said, 'It wisnae that, missus. It wis an auld fella that ordered the pie but he only ate the middle o' it. He couldnae get his gums through the crust bit.'

'He was eating with his gums?' the horrified lady exclaimed.

'That's right,' nodded the waitress, 'he'd came oot wi'oot his wallies.'

Sometimes we'd get off at Partick, which, like the Gorbals, was jam-packed with pubs and full of characters who were incredibly fast with their repartee. There were gangs of men drinking cheap wine on the corners. While they were having their street party, the characters would tell jokes and stories. We had just come out of the underground station when we spotted one character, wearing a bunnet, scarf and raincoat, holding court at a street corner, handing round a bottle of Four Crown wine. Two large policemen appeared on the scene and told the man off for drinking wine in the street. As quick as a flash, he replied, 'Officer, Ah thought it was wisnae an offence tae drink Four Crown in the *Royal* Borough o' Partick.'

When we got back to the Gorbals, we regaled some of the men with our adventures on the subway. One of them said that years before he had come up with the idea of doing a subway pub crawl around Glasgow, having a pint at a pub near each stop. He and his pals started at the Mally Arms near the Bridge Street station and, after 14 stops, they ended up in quite a state back where they started.

An alternative way to get around Glasgow was to take a ferry across the Clyde. We often crossed the river on the Finnieston Ferry or the Govan Ferry. At times, the Clyde looked beautiful, with its boats, cranes and warehouses, and it gave meaning to the saying, 'The Clyde made Glasgow but Glasgow made the Clyde.' For wandering bevy merchants, the ferries were handy to get to public houses across the river and there wasn't the problem of traffic jams. As soon as they got off the ferry, there was a pub across the road, which they scurried into for a pint.

At the height of my dogging days, I managed the odd cameo appearance at school and the headmaster called me in one day for a 'man-to-man discussion'. He said, 'We've given up trying to educate you, as it seems you are not happy with this school or the type of education we are trying to provide. So I'll tell you what, we'll do a deal. We've both got to cover ourselves, so as long as you register first thing in the morning, you can do what you fancy after that. Go and wander the hills if you want to.' I shook his hand on the deal but later found out that in reality the passport to wander the hills made life extremely boring, so much so that it could only lead to trouble. My father had never hit me, and it was my mother who was the only one to come up with any form of discipline. She warned me that the way I was carrying on I would end up in approved school.

I had a stand-up argument with her one day and I was extremely cheeky. The streets had made me too fast with my mouth. Next minute, she picked up her purse, which was loaded with change, and hurled it across the room at me. Bang! It was like a missile. It burst my nose straight away and there was blood everywhere. I ran out of the house pursued by my mother but quickly lost her in the labyrinth of the streets. I ran off to 'the shows', the carnival in the Glasgow Green. Not long afterwards, my mother turned up looking really sorry for herself. She said, 'Ah didnae mean to dae that, son. I was just annoyed that ye've no been going tae school. Come back hame and I'll make ye yir tea.' As I walked with my mother from the Glasgow Green to the Gorbals, I thought it had been a perfect end to a rather imperfect day. My dogging days were over.

Chapter 27

SERIOUS STUFF

During my teens, the Gorbals was disappearing before my very eyes. Tenements, pubs, shops, schools, playgrounds – you name it – were being torn down every day and the places we hung around simply vanished. A gang of us resorted to going over the bridge to the town, the city centre, to seek fun and adventure. The Central Station was, like the old Gorbals Cross, a place where you'd meet a wide variety of characters. There were rogues, thieves, pimps, prostitutes, muggers, all hanging about. The prostitutes got talking to men in the station and then offered them their services in one of the many dark lanes near by. As a result, there were lots of creepy-looking individuals lingering around trying to pick them up. I thought some of them were like characters from the television series *The Twilight Zone*.

We had a gang of more than 20 young guys all aged between 14 and 18 and we felt unconquerable. The Tiny Cumbie motto (apart from 'Tiny Cumbie ya bass!') was just like that of the three musketeers – 'one for all and all for one' – meaning that if any one of us got into a fight or any other trouble we would stand by each other. A few street girls approached us saying they were being pestered by some of the creepy men, so the gang decided to mug them. A girl led the punter into West Regent Lane and then a bunch of us hooligans appeared shouting and bawling: 'Gie's yir money or we'll dae ye in.' The men were too terrified to argue and handed their cash over. They never went to the police to make a complaint because of the situation they had put themselves in. Anyway, the other gang members and I reckoned we were doing a better job than the police in clearing up all the degenerates who hung around the Central Station like a bad smell. One night, a punter decided to put up a fight. But he had no chance against 20 of the gang. He got a bit of a kicking but still managed to make a getaway. A few months later, I was interested to hear a local politician talking quite eloquently on TV about how Glasgow's city centre had a growing crime and vice problem. He urged the authorities to act immediately with draconian measures to clear up the human filth that littered the streets. 'We have prostitutes, various criminals and razor gangs hanging about. It's time we came down on them like a ton of bricks,' he said. He also waxed lyrical about the virtues of family life in Scotland. His no-nonsense anti-vice and crime message surprised me because it was the same fellow we'd caught with the prostitute in the back lane a few months before.

As the gang stood in the station one night, a man and a woman with Irish accents approached us and began chatting away. The woman was wearing a beret and a dark leather coat. She told us with the nicest smile imaginable that both of them were in Sinn Fein and would we like to join? The message was clear – they were recruiting young Glasgow guys like us to be sleepers for the IRA. I gave the woman a false name and address because I did not want to get involved with the IRA – it was bad enough being connected with the Cumbie.

But the Sinn Fein couple were more interested in and more comfortable with boys whose parents were Irish anyway. My pal Chris told them that both his parents were from Ireland and their faces lit up. The woman said, 'That's grand, Chris, you're the sort of fella we're lookin' for. By the sounds of it you have a good Irish background.' Much to my surprise, Chris even gave them his real name and address.

A few weeks later, he was sitting at home when there was a knock on the door. It was the Sinn Fein couple. They introduced themselves to his parents and said they would like to have a private word with 'the boy'. They told Chris that they wanted him to join Sinn Fein and go on a training course in Ireland, which would last up to a year. Chris said he would think about it and they gave him their telephone number before they left. But Chris's father warned him not to get involved. Chris later phoned them saying he agreed with the principles of Sinn Fein, the IRA and a free Ireland but felt he was not yet ready to join.

One of the Gorbals boys did join and disappeared to Ireland for a while. When he came back after a year away, he was talking in an Irish accent and had plenty of money on him. 'I'm workin' to further the cause, and the struggle will never be over until there is a united and free Ireland. Our day will come!' he told me. I thought his patter had changed dramatically since he went to the Emerald Isle. The next time I saw him was when his face was splashed all over the papers. They said he was an active IRA terrorist arrested in connection with a series of bombings and other Republican attacks in London. It was strange thinking afterwards that if the boys and I had taken up their offer we could have become IRA terrorists. How would our lives have turned out?

One of the Gorbals gang ended up in Ireland for a different reason. Big Freddy really had a terrible life. He came from a broken home and had never known who his father was. His mother had a bit of a drink problem and also had a series of boyfriends whom Freddy was forced to call 'uncle'. Freddy ended up living with a nutty auntie and her sister in the Gorbals. Because of all his built-up aggression and the torture he had experienced in his life, he became one of the wildest boys around. He was always in fights and stealing. One night, at the age of 17, he got into an argument with a couple of guys at Gorbals Cross and got slashed. The scar on his face wasn't that big but it affected Freddy's self-esteem and he became downhearted about it. One day, he walked into the army recruiting office and signed up for ten years. In those days, it was as simple as that to sign your life away. He soon got his papers and rail travel warrant and was to undergo training in Pirbright, Surrey. A gang of us walked with him from the Gorbals into the city centre towards the Central Station. Before he got on the train and bade us farewell, he told us, 'The Army is ma last chance tae get away fae aw the trouble in the Gorbals. It'll keep me away fae the place and I'll learn a trade. It's ma only chance to stay out of jail and lead a normal life.'

A few weeks later, Freddy sent us a letter about his adventures in the army, about how fit he had got, how disciplined he was and how his life had changed for the better. About a year later he told us in another missive that he had been posted to Northern Ireland where he had met a beautiful Irish girl and got married. All of a sudden the letters dried up and we just presumed that Freddy had settled into a life in the army and marital bliss.

But some time later, one of the boys got a letter from Freddy in military prison. He had got back to his marital home and found his wife in bed with another man. He promptly blasted the guy with his rifle and subsequently received life imprisonment. So much for escaping from the Gorbals and keeping out of trouble. Alex summed it up perfectly: 'Freddy thought the army wid change him but it didnae. Ye can send a cabbage aroon' the world but it'll still come back a cabbage.'

As I have explained, when we were teenagers, most of the emphasis was on experimenting with drink rather than drugs. In fact, drugs were frowned upon as being a bit of a hippy thing so we rarely saw anyone hooked on drugs. Our views changed a bit, though, when a pretty hippy girl called Morag arrived in the Gorbals from Oban on the west coast. Her grandfather was one of the Highlanders who had come to the area during the 1950s. He wasn't that well and she had turned up to look after him. She was planning to study art and compared to the wild Gorbals herries she seemed comparatively demure, well spoken and sophisticated. While her grandfather was away having a hospital check-up she invited us up to his tenement flat in Florence Street. We expected her to have a carry-out at the house but she told us, 'I hate alcohol. I much prefer this, it's good stuff and you can get as high as a kite.' Morag then showed us a quantity of marijuana she had brought with her and asked if we wanted a joint to try it out. She rolled this enormous-looking cigarette and we all agreed that a couple of puffs could do no harm. I had never smoked anything before, not even a cigarette, and after a couple of puffs I felt extremely green. But after I got over the initial reaction, I found myself laughing hysterically at the smallest and stupidest of things. I just thought that marijuana was different from wine; at least you didn't get violent with it.

Another couple of puffs followed and when I went back to my house I found myself extremely interested in the floral-patterned wallpaper. I began to wonder how they managed to get all the different-coloured flowers into the wallpaper and was amazed at the design. I pointed this out to my pal Albert after what I thought was about two minutes but he laughed and told me that I had been staring at the wallpaper for more than an hour. If this was being stoned, I did not like it.

Meanwhile, one of the boys announced that he had discovered something better than alcohol: magic mushrooms. He said they would give us a nice feeling. I thought that as they were only mushrooms they must be all right. I swallowed a tablespoonful of the dried magic mushrooms and nothing happened straight away. So, after half an hour, I bade him farewell, declaring the mushrooms to be a load of rubbish. An hour later, I was sitting in a café talking to another member of the Cumbie when all of a sudden his face began to melt and drip like a candle onto the table.

I burst out laughing and said, 'Your face is melting ontae the table, ya big candle ye!' He gave me a puzzled look then worked out I was on drugs. He bade me a swift farewell, saying, 'Ah'm offski, you're aff yir heid the day.' After he left, a clock on the wall turned into a fat man's face with a handlebar moustache and he began grinning at me. Next minute, a phone rang and I declared that it was a murderer admitting that he had committed the crime. The other customers began looking at me as if I was insane. Then, all of a sudden, some of them turned into what looked like monsters. I rushed out of the place and headed to bed, trying to avoid this trip. But even when I shut my eyes, the inside of my eyelids became chessboards and it took what seemed like ages for the effects to wear off. After that, I vowed never to touch drugs again; it was a period of temporary insanity that I wanted to forget about. I just thought, 'Give me a pint of beer any day.'

But even that could lead to trouble. A few weeks before my fifteenth birthday, we were all sitting in a local pub knocking back pints of lager. There were about ten of us, all under the legal age, but the owner of this particular pub turned a blind eye to us drinking there. He was a weak-willed character who spent all of his takings propping up his betting and drink problems. And because he had become abusive to his regular customers, he had lost the majority of his trade, so he was quite happy to let youngsters like us into his pub – at least it helped fill his till.

But the older regulars who still hung around took exception to us drinking there and one of them, whom we suspected of being a grass, tipped off the police. We were all laughing and joking in the corner when two plain-clothes cops came in and told us to leave our drinks where they were. They then took our names, addresses and ages. I was sitting at the end of the table near the door so I grabbed my chance and ran out of the pub.

One of the policemen chased me a few yards but then gave up and went back into the pub to his captive audience. I ran back to my tenement in Crown Street safely but a bit out of breath. Lying on my bed, I congratulated myself on escaping from them. But then it dawned on me – I had given the police officer my real name and address before fleeing. The alcohol had clouded my judgement.

For a few days, nothing happened and I thought I was in the clear. But when I came back from school one day, there was a note behind the door from the Glasgow Police. It said they had called and demanded that I turn up for an interview over allegations of underage drinking. The date I was due to see them took me aback: it was 19 December, my 15th birthday. For a few days, I tossed and turned, barely able to sleep for worrying over my fate. But the night before the interview, I came up with a plan to deal with the situation.

The next day, I put Brylcreem on my hair and combed it back, giving it a thick, dark greasy appearance, in contrast to my normal fine, fair hair. I then put on an old pair of my father's heavy-rimmed glasses, dressed in one of my pal's school blazers and put on a white shirt and an old school tie. I looked at myself in the mirror and laughed, because I looked nothing like the boy who had been caught in the pub.

My father came with me to the police headquarters near Glasgow Cross. He said to me, laughing, 'Ah never thought I'd have a respectable schoolboy as a son! Ah'll tell ye whit, ye certainly look the part the way ye've done yirsel' up.' When I went into the interview room, the two policemen who had caught me in the pub were there. They looked a bit puzzled at seeing me but tried to hide the fact by asking me lots of questions about why I was in the pub and how often I went there. I replied, in a voice that sounded nothing like my real one, 'I don't know what you're talking about, somebody must have given you a false name – mine.' It was call-my-bluff time. Then one of the officers used the old psychological ploy of staring me straight in the eye, asking me more questions.

But I stared right back at him for what seemed like an eternity and, much to my surprise, he blinked. It was eyeball to eyeball and the other guy had just blinked! They agreed it must have been someone else and let me go. I got back to my house and washed the Brylcreem out of my hair, discarding the blazer and tie in the process.

It was my fifteenth birthday and as I blew out the candles on my two and sixpence birthday cake, I felt on top of the world.

Chapter 28

WHO MURDERED
THE GORBALS?

A few weeks after that brush with the law, I left school and I felt no great sentiment about it. I had to get myself into a trade and within a month I landed employment as a commis chef at the Royal Scottish Automobile Club in Glasgow's Blythswood Square. My father had worked there years before and was on friendly terms with the head chef, so the job was quite easy to get. The wages, however, weren't particularly good – £6.50 a week – and sometimes I had to do shifts of up to 12 hours. But my parents said it would keep me away from the Gorbals gangs and out of trouble, and they proved to be spot on. The other teenagers I'd been hanging around with were getting arrested for all sorts of different and outlandish reasons.

Meanwhile, the area continued to be reduced to a pile of rubble. It was a case of half of the Gorbals was there and half of it wasn't. I'd go to work in the morning and by the evening when I came back a pub or shop which had been there for years and years and had been a vital part of the community would have been knocked down. The men and women who had worked there, with all their stories and patter, disappeared as well. Slowly but surely, the old Gorbals was being well and truly killed off.

I witnessed people, mostly elderly folk, crying in the streets as their tenements were flattened to the ground. 'There goes ma life,' shouted one pensioner as a huge bulldozer destroyed her tenement in Florence Street. It was an extraordinary experience watching the Gorbals crumbling like some giant dinosaur that had outlived its time. Most days there were removal vans everywhere loading up furniture, to be decanted all over the city and in the new towns throughout Scotland. A couple who had run a corner shop in Hospital Street stood and looked on aghast as their tenement and shop were flattened. The tearful woman said to me, 'They've destroyed ma tenement where ma family stayed for mair than three generations and they've smashed up ma wee shop. Now they've moved us tae a council housing estate and there's nothin' there. Big businessmen are making plenty of money out of this and we're only pawns in their game. The Gorbals means nothing tae them, it's aw aboot them makin' a bloody profit.'

We found it hard to understand how they could knock down a place which had its own distinctive history and way of life. Numerous majestic, magnificent buildings were torn down without a second thought. Indeed some of the buildings were comparable to the ones which still stand in highly regarded parts of Edinburgh. It didn't take a mastermind to wonder why they didn't just modernise and refurbish many of these awe-inspiring old buildings and keep the Gorbals flourishing. The politicians blamed it on disintegration, overcrowding and inadequate sanitation, which was widespread. But to the locals, the point of view at the time was that because of the bad press the area had got over the years, initially sparked by *No Mean City*, Glasgow Corporation wanted to wipe the area off the face of the earth, kill it off completely. Post-war housing legislation was supposed to create a land fit for heroes. But our argument was that new council estates like Castlemilk didn't exactly fit the bill.

Glasgow Corporation's redevelopment schedule marked down 29 neighbourhoods as Comprehensive Development Areas. The Gorbals was to be the first casualty. Hundreds of thousands of people from the Gorbals and the rest of Glasgow became merely overspill statistics. People who were being moved out to the new housing schemes and towns took exception to it because they yearned for the disappearing Gorbals way of life.

An examination of the old *Evening Citizen* through the years traces the annihilation of the Gorbals, which really started in earnest years before, in 1961. At that time, it was reported that there were 98.9

persons per acre in the Gorbals, compared with 18.7 in Kelvinside and 11.9 in Pollokshields. Distinguished Scottish architect Basil Spence was given a £1.3 million contract by Glasgow Corporation for a new-look Gorbals. For inspiration, Spence made a special excursion to Marseilles to see Le Corbusier's Unité d'Habitation. A Corporation spokesman said the French buildings might be the solution to the blueprint of the new Gorbals. Afterwards, it was announced that the Queen was to pay a visit to the 'exciting new development', seeing both the best and worst of the neighbourhood. War veteran Johnny Shannon, aged 75, who had just been moved from his crumbling tenement in Thistle Street to a brand-new flat near by, said, 'I'd love to meet the Queen and show her around my spanking new apartment.'

The *Evening Times* and *Citizen* reported that when they were told that the Queen was likely to visit their condemned tenement in Sandyfaulds Street, the local women scrubbed out their closes and removed graffiti from walls. The clearly taken-aback monarch looked round Catherine Dempsey's tiny single end and enquired, 'Is this all you have?' Mrs Dempsey explained afterwards, 'I think she was amazed at the smallness of the house.' In the same street, Mrs Betty Meek told the Duke of Edinburgh to be careful, as the floorboards were disintegrating and might give way. The Duke glanced around Mrs Jean Percy's living room in her contemporary flat in Commercial Row and asked if it was 'all on the never-never'. Mrs Percy proudly replied that everything had been paid for. One of the royal party, a condescending young equerry, was reported in the papers to have said, 'We'd all heard so much about the Gorbals. We imagined we'd be walking through three feet of mud.'

The Queen was also shown a model of Basil Spence's vision for the pioneering sky-scraping flats. Spence maintained that each and every family would have a balcony with a garden, even at the twenty-storey level. Such claims were met with much scepticism by local people. They were right: in later years these promised gardens in the sky did not materialise.

In 1962, Tory Secretary of State for Scotland Michael Noble proudly opened a new high-rise block in Adelphi Street and Norway's King Olav, on a state visit to Scotland, was keen to witness the new Gorbals emerging from the old. The king looked on without emotion as an aged Glasgow slum was torn down just a few yards away from Basil Spence's newborn concrete blocks. As buildings were being demolished all over the place, Hugh Fyfe, aged 82, who had fought in the First World War, refused to move out of his slum tenement in Mathieson Street, which was due to be knocked down, saying, 'The Jerries couldn't scare me and neither can the Corporation.'

In 1964, the *Evening Citizen* reported that the development of medium- and high-rise flats between Ballater Street and the River Clyde was complete. The new development included the four tower blocks of Commercial Court and Waddell Court. Jewish community leaders set up a house-purchase scheme to help the remaining Jews in the area to move to parts of the city with an established Jewish population. In the same year, Gorbals families who had been rehoused in the new flats petitioned Glasgow Corporation for a ban on public houses in the area, backed by local councillor, later to become MP, Frank McElhone. A mere nine taverns were to take the place of the previous forty-seven in the immediate area.

Many housewives suffered from a new syndrome, the skyscraper blues, and there were suggestions that more local playgrounds and nurseries be built. Comedians had a field day. There was a story going round about the housewife who suffered particularly badly from the skyscraper blues. Depressed, she struck and killed her husband with a frying pan. She rushed from her tenth-storey flat and shouted to a neighbour, 'Oh, Betty, Ah've jist killed ma man! Ah'll get life!' Betty, a lady of much astuteness, urged her to keep composed. 'Jist go back tae the hoose,' she advised, 'put a wet chamois cloth in his haun an' fling him oot the windae.'

But some high-rise inhabitants cherished their new modern lifestyle. Mother of three Mrs Rogers moved into a top-floor flat in a multi-storey in Rutherglen Road. She told the *Evening Citizen*, 'I wouldn't change my new flat for anything. The weans are not allowed to play in the corridors but they have a swing park opposite the building. At first I didn't want to live so high, but we have such a marvellous view. I wouldn't swap my flat now for one any lower.'

When the tenements were being knocked down, hundreds lay unoccupied for months before demolition. Alex, Albert, Chris and I surmised there had to be treasure hidden underneath the floorboards, as older Gorbals people were real hoarders. A mob of us equipped ourselves with hammers, saws, pliers and a jemmy and then proceeded to take up the floorboards in every empty house we visited. It was incredible what we found: wedding rings, necklaces, pipes, old bank notes, coins, watches and photographs. It was almost as good as being an archaeologist digging for remnants from ancient times. One day, Chris lifted up a floorboard and shouted, 'Oh, ya beauty, ye!' He'd found an old war medal in mint condition. He was so overjoyed he rushed with the medal up to a merchant in the Barras who promptly gave him £20 for it. But when we told one Gorbals war veteran about the medal and the amount of money it had fetched, he blasted us: 'Ye've been robbed. That medal wis worth at least a couple o' hundred pounds – at least!' Many of the bits and pieces must have been of great sentimental value to their owners, who had probably passed away. The hoards were like miniature time capsules, old hidden secrets underneath the floorboards.

Many of the former tenement dwellers moved into the Basil Spence-designed high-rise blocks at Queen Elizabeth Square, which eventually accommodated 10,000 people, much fewer than the 27,000 who had lived in the tenements there before. The new buildings also changed the atmosphere of the community. People complained that the flats had left them living in 'streets in the sky', with the friendliness and intimacy of the old days gone forever. Spence said that the multi-storey flats would 'look like a great ship in full sail' on washdays. People found this observation patronising, as he must have been thinking of the Gorbals many years before, when people hung their washing out in the Glasgow Green. In 1966, downhearted residents told the papers that vandalism was rife, blaming it on the lack of amenities for young people. They had given each of the blocks a new nickname: Alcatraz for Block A, Barlinnie for Block B and Sing Sing for Block C. The Corporation hit back by saying the scheme was still very much at an experimental stage.

One local man I knew moved to the flats but became so depressed he threw himself off a tenth-storey balcony, and several other former tenement dwellers chose the same way out. They were just statistics that would be forgotten, the obscure casualties of the Gorbals' demise.

Some were deeply apprehensive about being moved to the housing estate of Castlemilk. 'For God's sake,' one of the bunnet brigade complained to me, 'there's no even wan pub in Castlemilk. The place is a desert wi windaes. It's worse than bein' in jail. At least in Barlinnie ye've got a bit o' atmosphere.'

One amazing sight was watching the poorer folk who could not afford a removal van do a flitting. They often loaded up all their possessions on a cumbersome old pram. It would be weighed down with smaller items like kettles, irons and small ornaments on the inside, while a wardrobe or armchair would be placed on top and the pram would wobble along the road with its heavy load.

In the 1930s, the Gorbals had an official population of 90,000 and was served by 1,000 shops and 130 pubs. But by the end of the 1960s, that would all have changed, with most of them completely destroyed in the most uncaring and malicious fashion. Locals said that the obliteration of the old Gorbals reminded them of the German bombing raids during the war; the difference was the Germans failed – Glasgow Corporation didn't.

But, of course, the Gorbals children still played, adapting to their new lives in the high-rise flats. In the old days, if they were out playing in the streets, their mammy would throw them a bread-and-jam piece out of the window – one of the earliest forms of fast food. The piece would be wrapped up in paper in case it hit the ground. But you couldn't exactly do that out of a high-rise block, could you? This problem led to the composition of the famous 'Jeely Piece Song' by folk singer Adam NcNaughton, which drew attention to the drawbacks involved in throwing a sandwich out of a 20-storey flat.

Many families who were angry or depressed about emigrating to the new council estates didn't stay long. They thought up every conceivable plan to return to what remained of the Gorbals. This included giving bungs to factors who still had buildings standing which would not be redeveloped for several years.

One of my older neighbours, Mr McDonald, was offered a brand spanking new house in Castlemilk but paid a factor his life savings – a hundred pounds – to continue his existence in the Gorbals. He said to me, 'Whit good is a hundred pounds in Castlemilk when ye're no happy? I'd rather be skint and be back in what is left o' the Gorbals. This is where Ah'm happiest.'

One of the first boys I knew to be decanted from the Gorbals was my pal Albert. His family, who were very poor, lived across the road in Crown Street. He often went for messages for my mother and preferred to stay most of the time in my house rather than his own. He had an unhappy childhood, living in appalling conditions. The fact that his father was a big boozer did not make life any better for him. With a mop of orange hair and a cheeky smile, Albert tried to make the best of his dire circumstances and put an optimistic light on the darkest of situations. In the early days, we'd wandered the streets and back courts of the Gorbals and even trailed up and down closes selling bags of sticks from the local sawmill, saying, 'Dae ye want tae buy a bag o' sticks fur wan and six?'

I met Albert one day in Crown Street and he looked downhearted. He said his family were being decanted to a new town near Edinburgh and he was heartbroken to leave. 'They might as well be sending us tae the moon,' he complained. 'Ah don't know anybody there but they promised ma da and maw work in wan of the new factories wi a new-fangled hoose tae go wi it. Ah don't want tae go, aw ma pals are here in Glesga, no in Edinburgh.' At first, I missed him dreadfully but a few weeks later he turned up at our door looking bedraggled. He was crying, saying he could not stand living in the new town any longer and had run away. In fact, he had actually walked all the way to the Gorbals, over two days, sleeping in fields. My mother made him a slap-up meal and said he could stay as long as he wanted. She was so moved by his plight she even offered to adopt him. For several days, Albert lived happily with us as one of the household. But one night, the police arrived at our door, said Albert had been reported missing from his new dwelling and took him away.

From then on, he tried to visit us whenever he could. One day, he turned up with two of his new friends, wild-looking youths who had strong Midlothian accents. We had a meander through the streets, with lots of patter and banter flowing, and Albert was proud to introduce me to the boys as his best buddy in the Gorbals. We had a good laugh but as I walked them to the bus stop, alarm bells began ringing in my head when one of the youths started boasting that he was a car thief. I waved them goodbye as their bus pulled out of Glasgow Cross, then I walked back over the bridge to Crown Street, reflecting on what the boy had said and the way he'd said it. My overall impression was it had been a bit disturbing, even something of a bad omen.

A few weeks later, I walked into the house and there was complete silence. My mother looked as though she had been crying. She then walked over and said, 'Look at this,' giving me a copy of the day's newspaper. Albert's photograph stared out from the page and the headline said: 'Three boys killed in car crash'. I let out a howl; it was like the scream of a wild animal. I filled the house with my grief, realising I would never see Albert again.

We managed to stay in Crown Street until the very last moment – 1972. By that time, most of the flats in my tenement were boarded up and the local delinquents had started to decorate the close with grafitti. The reason we stayed so long was because my father had turned down several offers of accommodation from the Corporation. 'There's no way Ah'm going tae somewhere like Castlemilk,' he maintained. He said he had a pal in the Corporation and would not hesitate to resorting to bribing an official 'tae get a decent hoose in a decent area'. A few weeks later, out of the blue, he was alerted that a small Corporation flat in Shawlands, a comparatively middle-class area, had come up. We went to see it and although it had even less living space than our Gorbals tenement, he snapped it up immediately. 'We've done oor time in the Gorbals,' he said. 'Noo it's time tae move onwards and upwards.'

A week later, the flitting van parked outside of our close and as we loaded our furniture into it, remaining neighbours and friends came up to say their final farewells. 'It's the end of an era,' Mrs Carey, my former neighbour, said. 'They've managed tae destroy one of the greatest places on earth.' She pointed to the new high-rise flats. 'Look at them – how dae they expect people tae live there? They're no real hooses, they're boxes.'

As I walked through the Gorbals for the last time, all my boyhood memories came flooding back: the hilarity, the tears, the characters, the patter, the extraordinary escapade of being a part of the old community. Someone had even spray-painted a wall with giant letters asking 'Who murdered the Gorbals?' On seeing the message, I felt angry and bitter at first. But then I could not stop laughing, realising that someone out there actually had the cheek to buy an aerosol can and deface a wall with such a significant question. Now that was the real old Gorbals spirit.

As our flitting van pulled out, heading towards a brand-new life, my father summed it up: 'Cheerio, Gorbals, the show is all over. But there's an auld saying: if you stay long enough in a place, you become that place!'

POSTSCRIPT

In 2007, I am walking past what is left of Gorbals Cross and everything looks grey and depressing. Not much life here, just the monotony of traffic on its way to and from the city centre. I suddenly realise that it is 40 years since we all marched through here when Celtic won the European Cup. Four decades since I left St Bonaventure's. That thought, plus an icy wind blowing across the Clyde, puts a chill through my bones as ghostly images of the past come back to haunt me. For a minute, I get paranoid and question whether all these long-ago things really happened to me and indeed if the old Gorbals ever actually existed at all.

Near Gorbals Cross, a large mosque has been built and the new Sheriff Court stands not far away – this is certainly not the area as I remember it. I begin to tremble; for me, this is a form of future shock. The Citizens' Theatre is still standing but looks like a pale shadow of its former self. I feel like Winston Smith in *1984*, searching for anything that will be a reminder of the past.

My spirits are low until I arrive at the corner of what is left of Gorbals Street and Cumberland Street and discover the Brazen Head pub. There is a rousing sing-song going on and the place is swathed in green. After a few minutes, I begin to recognise a series of old Gorbals faces. I am introduced to a half-cut fellow whom I went to school with all those years ago. He looks me up and down and laughs saying, 'Ye've changed yir style since Ah last saw ye, and ye've put on a bit o' weight.' I find this remark pretty funny, as the last time I saw him was probably 30 years ago. Did he really expect me to still look like a teenager after all this time? In the pub, there is an air of wildness and drunkenness, just like the old days. Men stand around telling stories and jokes, and amidst the banter there is raucous laughter all round; this cheers me up no end, as it is pleasing to see the old characters have still survived, even if only in ever-decreasing numbers. 'Hey, Ah remember you, you were in the Cumbie the same time as me,' says a scarred guy at the bar. He then gives out a cry of 'Cumbie ya bass!' This has me and the rest of the place smiling in bemusement. Standing near by, an old Gorbals diehard is telling a series of quickfire jokes: 'A penguin walks intae a bar and says tae the barman, "Has ma faither been in?" And the barman says, "Whit dis he look like?"'

After half an hour, I am engrossed in the atmosphere and I am amazed how quickly people drink. I buy a rather large round at the bar and no sooner has it been shifted than another round appears, then another and another. Later, I find myself in the Lauriston pub in Eglinton Street. It's about the last genuine Gorbals pub and the decor hasn't changed in years. My grandfather and father once drank there.

A series of ageing fly men and patter merchants come in. One announces that he is going outside for a cigarette; the smoking ban has recently come in. 'Whit a palaver,' he says to me, blowing plumes of smoke down Eglinton Street. 'That Scottish Parliament are making up aw these new laws and they couldnae even run a brothel wi two beds. In the old days ye could sit doon and hiv a fag wi yir pint. But times have changed. We noo live in an age where it's legal for a man tae marry a man but ye cannae have a fag indoors!'

I smiled, just thinking that the old Gorbals may be dead but the patter and the people still live on.

Appendix

FAMOUS GORBALS PEOPLE

ARTHUR BLISS

The opera *Miracle in the Gorbals* was composed by Arthur Bliss in 1943 and was first produced by the Sadler's Wells Ballet at the Prince's Theatre, London, on 26 October 1944. It was set in the Gorbals slums and is about a prostitute who drowns herself but is later brought back to life by a stranger. The plot leads to a murder that is a metaphor for the evils that existed in the Gorbals at that time.

JIMMY BOYLE

Former Cumbie gang member Jimmy Boyle was convicted of murdering Babs Rooney in 1967. He still maintains his innocence and has been at the forefront of demanding changes to the Scottish Prison Service. His autobiography, *A Sense of Freedom*, was written in Barlinnie's special unit, which focused on the rehabilitation of offenders, and was adapted for the screen in 1979. After his release, he became a successful sculptor and charity worker, and he now splits his time between an Edinburgh mansion and a Marrakesh villa.

PAT CRERAND

Pat Crerand was brought up single-handedly by his mother, his father having being killed in an air raid during the Second World War. He was born in Thistle Street and his family later lived for many years in Crown Street. He was educated at St Luke's Primary School and Holyrood

Secondary. He learned his footballing skills in the back courts of the Gorbals and joined Celtic Boys Club. He left Celtic in 1965 and later won the European Cup with Manchester United in 1968.

TOMMY DOCHERTY

In 1972, Tommy Docherty took over the management of the Scotland football team. He gave the national side a new energy and pride. In twelve matches under his reign, Scotland won seven and lost only three, and those were against strong opposition: England, Brazil and Holland. He then took over at Manchester United, with whom he won the FA Cup in 1977, beating Liverpool 2–1. He is now a successful broadcaster and commentator.

RALPH GLASSER

In 1986, economist and psychologist Ralph Glasser published the first volume of his autobiography, *Growing Up in the Gorbals*. Born in 1916, he left school at 14 and worked as a presser in a rag-trade sweat shop before winning a scholarship to Oxford – and he cycled all the way from Glasgow, 400 miles, to get there. He has said of his childhood, 'Life was hard, but I have a lot to thank the Gorbals for. If you boil it down, growing up in the Gorbals meant that maturity was forced on you much earlier. There was little opportunity to indulge the usual childhood fantasies.'

ALEX HARVEY

In 1982, Alex Harvey died of a heart attack in Belgium on the eve of his 47th birthday while returning from a European tour with the Sensational Alex Harvey Band. The band has been described as 'heavy metal when the rest of rock and roll was still in the Stone Age'. Reared in Thistle Street, he made his name in the 1950s with the Alex Harvey Big Soul Band. After he married and moved into a single end in Crown Street, he gave a series of free gigs to raise funds to take local pensioners on bus trips. He went on to earn an estimated £15 million from pop music, helped by the Top 20 hits of the mid-1970s 'Delilah' and 'The Boston Tea Party'.

LORRAINE KELLY

The TV superstar lived in a single end in Ballater Street as a child, before moving to Bridgeton and later to East Kilbride, where she got her first job as a journalist on the *East Kilbride News*. When working as a researcher for BBC Scotland, she was told she would never appear in front of the camera unless she got rid of her Glasgow accent. She left the BBC and Glasgow for London to join the newly launched *TV-AM*, and a hugely successful career in daytime television followed. She became a household name with *GMTV*, which currently features her *LK Today* show. She has hosted *Have I Got News for You* and been a regular stand-in presenter on *This Morning*.

HELENA KENNEDY

Helena Kennedy QC is one of Britain's finest legal minds. She was born in the Gorbals and attended Holyrood Secondary School. She was called to the English Bar in 1972 and has been involved in a number of high-profile cases. She was made a peeress in 1997 and concentrates her energies on civil liberties. Her books include *Eve Was Framed* and *Just Law*.

BENNY LYNCH

One of Scotland's best boxers was Benny Lynch from Florence Street. His life was a classic roller-coaster story: one minute he was poor and unknown, the next he was rich and famous; but like many other Gorbals characters, his fortunes reversed after he turned to alcohol. He was the first Scottish boxer to win a world title, defeating Jackie Brown in Manchester and Small Montana in London to take the world flyweight title. He was still a young man when his career ended and he battled with alcohol for the rest of his short life.

ALEXANDER McARTHUR

Alexander McArthur was the co-author of the most controversial book written about the Gorbals – *No Mean City*. He was born in the Gorbals in 1901. As a baker in and out of work, he dreamed that one day he would become a famous writer and set about writing two novels, which were rejected by London publishers Longman's for being too rough around the edges. But his lurid descriptions of how tough life was living in the worst slums in Britain showed great potential. The publishers hired highly experienced journalist H. Kingsley Long to ghostwrite a novel with McArthur. The publication of *No Mean City* in 1935 caused a sensation. It was a huge bestseller and has remained in print ever since its publication but McArthur earned very little from it and achieved no further success. He was found dead in the Clyde in 1947.

FRANK McELHONE

In 1973, Granada TV's *World in Action* made a documentary about MP McElhone's Saturday surgery, which was held in a former bank in Gorbals Street. It was known throughout the Gorbals as 'Frank's Bank'. A spokesman for *World in Action* said, 'We didn't choose the Gorbals because of the usual *No Mean City* angle. We were there to praise Mr McElhone's work in his constituency and pose relevant questions. There is a strong political line to the programme and we will be asking why the Government cannot make sure every MP in the country has a surgery.' Following his election in 1969, McElhone campaigned relentlessly to have the old Gorbals knocked down and a new one built.

ROBERT McLEISH

McLeish made his name worldwide by writing *The Gorbals Story*. Born in 1912, he became a plumber but unemployment during the Depression led him to a writing career. His experiences in the area inspired him to write a play for the Glasgow Unity Theatre called *The Gorbals Story*. It was performed more than 600 times between 1946 and 1949, and was made into a film in 1950.

ALAN PINKERTON

The world's first private detective came from the Gorbals. Alan Pinkerton emigrated from the Gorbals to America, where, in 1850, he set up the detective agency Pinkerton's with its distinctive 'open eye' logo and the motto 'We Never Sleep'. The agency quickly became known as 'The Eye' and as a result all private detectives since have been known as private eyes. The agency tackled such cowboy outlaws as Butch Cassidy and the Sundance Kid and the Jesse James Gang. Pinkerton uncovered a plot to assassinate Abraham Lincoln and the two were great friends.

JOHNNY RAMENSKY

Johnny Ramensky was a legendary safe blower who was in and out of jails all of his life. Known as 'Gentle Johnny', he was parachuted behind enemy lines to work for the secret service during the Second World War. The son of a Lithuanian miner, he followed his father into the Lanarkshire pits where he learned to use explosives. After his father's death, the young Johnny moved with his mother and sisters to the Gorbals. At the time of his death in 1972 he had been sentenced to a total of 56 years in jail since his first conviction in 1917.

REVEREND GEOFF SHAW

In 1960, Reverend Shaw moved into a two-room-and-kitchen flat in a dilapidated tenement at 74 Cleland Street and soon local people were beating a path to his ever-open door. With the help of other members of the evangelical Gorbals Group, he helped halt evictions and organised rent strikes to force landlords to carry out essential repairs. Like fellow minister Cameron Peddie in the 1930s, he also worked with local juvenile delinquents.

JAMES STOKES

Private James Stokes, who served with the King's Shropshire Light Infantry, fought heroically during a battle with the Germans at Kervenheim in Holland in 1945. He captured 17 enemy soldiers, putting himself at grave risk and was mortally wounded. He had worked as a labourer in the Gorbals until he joined the army. He was just 29. Stokes was posthumously awarded the Victoria Cross. In a letter to Stokes' widow, Janet, who was living in a single end in Clyde Street with her young son and her brother and sister, his commanding officer wrote, 'His actions were those of a hero. I have never been so proud of anyone under my command.'

SIR ISAAC WOLFSON

Ex-Gorbals boy Sir Isaac Wolfson, who was born in a run-down tenement in Hospital Street, was made a Freeman of the City of Glasgow in 1971. He was the founder of Great Universal Stores, which became a billion-pound business worldwide. After leaving the Gorbals, he lived in a mansion in London's Portland Square but he admitted, 'I just lead a simple Glaswegian way of life.' By 1971, through the Wolfson Foundation, the tycoon had given away £17 million to groups and institutions all over the world.